Dealing with Failed States

With the ever-increasing interdependence across individuals, groups, international organizations, and nation-states an increasingly significant policy concern in the contemporary turbulent world of globalization is the question of state failure. There has been a growing academic interest in the determinants of state failure and an acute awareness across the international community of the need for dealing with issues of instability in states.

The contributors to this volume represent the most recent cutting edge approaches to state failure—looking at both conditions of conflict and economic development, dealing with the conceptualization, causes, and consequences of state failure, as well as policy-oriented analyses as to how state failure can be contained, reversed, or prevented. In order to deal fully with the phenomenon of state failure, investigators *must* be involved in a number of boundary-crossing activities. The contributors to this volume have addressed failed states through:

- multiple levels of analysis, assessing domestic and cross-border phenomena, internal and external conflict, domestic and international political economy;
- multiple disciplines and interdisciplinary approaches representing political science, sociology, and economics;
- various methodological approaches, including large-N empirical analyses, case studies, and simulations; and
- through both basic and applied research, drawing on the work of academics, IGOs, NGOs, and national governments.

This book was originally published as a special issue of *Conflict Management and Peace Science.*

Harvey Starr is the Dag Hammarskjold Professor in International Studies in the Department of Political Science at the University of South Carolina.

Dealing with Failed States

Crossing Analytic Boundaries

Edited by Harvey Starr

Routledge
Taylor & Francis Group
LONDON AND NEW YORK

First published 2009 by Routledge
2 Park Square, Milton Park, Abingdon, Oxon OX14 4RN
605 Third Avenue, New York, NY 10017

Routledge is an imprint of the Taylor & Francis Group, an informa business

Typeset in Times by Value Chain, India

British Library Cataloguing in Publication Data
A catalogue record for this book is available from the British Library

ISBN13: 978-0-415-48332-2 (hbk)

Contents

Acknowledgements

I would like to thank Glenn Palmer for suggesting that there would be great interest in an in-depth set of papers on failed states, and encouraging me to take on the task. Thanks also to the set of anonymous referees who blind-reviewed all of the submitted papers and provided useful comments for revision, and Steve Thompson of Routledge who helped to bring an edited volume on failed states to fruition. Finally, thanks to the contributors who not only provided such first-rate papers, but who also met the series of deadlines I set for them.

Harvey Starr
Department of Political Science
University of South Carolina
Columbia, South Carolina, USA

Introduction:

Dealing with Theory and Analysis of Failed States

HARVEY STARR

The phenomenon of "failed states" has drawn increasing attention as a significant issue in contemporary human security, due to the consequences of the collapse of order within a nation-state's legal boundaries for a state's people, its neighbors, and the international system as a whole. In general, these consequences involve the generation and spread of violent conflict as well as the proliferation of poverty, disease, and human dislocation. Thus, state failure is a phenomenon of incredible importance to the study of domestic and international politics in the contemporary world system, with consequences that have the potential to ramify for decades to come. The challenges of understanding state failure, as well as how to deal with it, have been taken up by political scientists, economists, and sociologists, among others. Within political science, it has drawn the attention of both students of International Relations as well as Comparative Politics. Within the subfield of International Relations, the study of state failure can, in turn, be placed within the areas of both conflict and political economy (which includes CPE and IPE).

As such, state failure is exemplary of the sort of contemporary political phenomena that "cross boundaries," and thereby requires scholars to cross a variety of analytic boundaries as well (see, for example, Starr 2006 or Werner et al. 2003). In a number of ways, academic scholars have come late to the area of failed states, following the lead of international organizations and agencies (such as the World Bank), or the governments of the major aid-providing states (for example, the U.S., UK, Canada). This tardiness rests in part on the problems of crossing boundaries, or the unwillingness to do so. As I have noted elsewhere (Starr 2006a, 1):

> Boundaries can indicate the limits of some set of phenomena; such simplification and specification can be valuable in the development of theory, concepts, and research design. However, boundaries too often loom as *barriers*, which can hinder how we think about phenomena, how we theorize about phenomena, and how we study the world about us. As scholars we must be conscious of artificial boundaries or barriers that constrain our thinking, and be just as conscious of finding ways to promote fertile theory and effective research design. In this sense we should think of the crossing boundaries approach *not* as some new theory or theoretical approach, but rather as a synthesizing device that helps us in organizing theory and research.

In order to deal fully with the analytical, empirical, and policy aspects of state failure, investigators must undertake a number of boundary-crossing activities and adopt a diverse

range of theoretical and methodological approaches. The authors here have addressed failed states through:
(a) Multiple levels of analysis, by assessing domestic and cross-border phenomena, internal and external conflict, and domestic and international political economy.
(b) Multiple disciplines and interdisciplinary approaches, representing political science, sociology, and economics, with much of the work discussed being informed by spatial approaches reflecting the work of geographers.
(c) Various methodological approaches, including large-*N* empirical analyses, case studies, "intelligent data analysis" and simulation.
(d) Analyses that have generated both basic and applied research, drawing on the work of academics, IGOs, NGOs, and national governments.

Within a broader discussion of the most recent and cutting edge approaches to state failure, these studies present investigations of the causes and consequences of state failure, as well as analyses that inform ways in which the effects of state failure can be predicted, contained or reversed. Thus, the following works report on both basic and applied research, summarizing much of our current understanding of state failure, and specifically addressing policy choices. I believe that they represent both genres of research well, and provide an important synthesis across them.

The two opening pieces, by Jack Goldstone and Robert Bates, ask why state failure occurs, and examine explanatory/causal factors (but see also Carment et al., Chapter 5). Goldstone, drawing on his past work on revolution and civil war, as well as his participation on the State Failure Task Force (now Political Instability Task Force, PITF), sets out an analytical framework for ordering the multiple possible causes– pathways– to state failure. He notes the need to find a "middle ground" between the view that each state failure is unique and more general propositions about the causes of state failure; Iqbal and Starr (Chapter 3), and Carment et al. (Chapter 5), also note that large-N studies need to be supplemented by comparative case studies. Goldstone presents five major pathways, or stylized scenarios, which "may combine in various sequences": escalation of communal group conflicts; state predation; regional or guerrilla rebellion; democratic collapse; succession or reform crisis in authoritarian states.

While Goldstone identifies "effectiveness" and "legitimacy" as the factors underlying "stability," Bates draws on his previous formal work to identify conditions necessary for "political order" in his investigation of why things fell apart in late-20th century Africa. Bates argues that political order involves two types of choices: the choice of government leaders to act as warlords or as statesmen, and the choice of citizens to arm or live peacefully. Note that the government choice to act as warlords equals "predation" which is one of Goldstone's pathways. The basic explanatory framework used in Chapter 5 by Carment et al. to explain state "fragility" is ALC – autonomy, legitimacy, capacity.

In Chapter 3, Iqbal and Starr address the question of why state failure is important. Starting from the position of many IGOs and aid donors that state failure endangers other actors or the international system itself, they look at the effects of state failure within a diffusion framework to assess "the negative effects of state collapse." They find that political instability and unrest, civil war, and international war do diffuse from failed states, but *not state failure itself*. Their discussion of this finding links directly to the more explicit policy orientation of Chauvet and Collier (Chapter 4), and Carment et al. (Chapter 5), who address how to deal with state failure. The policy conclusions/recommendations of these latter two chapters on the use of aid (and other intervention) may help explain why state collapse itself does not appear to be contagious. Note also that Iqbal and Starr (Chapter 3) look at the political and security effects of state collapse, while Chauvet and Collier (Chapter 4) look specifically at economic factors and effects within a policy context– two areas Iqbal and Starr highlight for future research, especially as possible strategies through which IGOs and donor states might forestall the diffusion of collapse.

Looking at "failing states," Chauvet and Collier in Chapter 4 search for the conditions that underlie "sustained policy turnarounds" by which states can reverse the state failure process. They conclude that "appropriate donor intervention can radically shorten state failure". The reform required for a turnaround is found actually to be retarded by financial aid, especially what they term "windfall external finance." On the other hand, technical assistance is found to be more effective in promoting a turnaround. In Chapter 5, David Carment, Yiagadeesen Samy, and Stewart Prest also focus on assistance policy, and the relationship between development assistance and state fragility. Like Chauvet and Collier, they question blanket views of development aid, and stress that more work needs to be done on how different types of aid provided under different conditions, will affect state fragility.

The final chapters also introduce new methodological tools to the study of failed states. Englehart and Simon, looking at state failure as the result of dynamic processes (rather than as an on-off onset by which it is present or absent), set out a simulation that captures one path to state failure—a self-destructive internal dynamic of state predation that interacts with external conditions and strategic choices. Lambach and Gamberger introduce a new methodology based on machine learning tools in order to re-analyze the data generated by the Political Instability task force. This method allows for the "discovery" of subgroups, and ultimately models which identify countries susceptible to political instability.

The careful reader may have been struck by perhaps the most problematic aspect of the study of state failure and of the extant literature – issues of conceptualization, definition, and measurement.[1] The chapters in this book reflect the heterogeneity (diverse and varied state) of definition and measurement regarding "state failure" or "failed states." In brief, Goldstone defines state failure in terms of "stability," while Bates is interested in "political order." For Bates, "the mark of state failure is the government's loss of its monopoly over the means of coercion." Iqbal and Starr, after raising the issue of conceptual and definitional disarray, see state failure as "collapse," using the POLITY code of "the complete collapse of central political authority." Chauvet and Collier are interested in "failing" states, based on "inadequate performance of socio-economic functions." Thus, they study low-income countries with extremely weak economic policies, institutions and governance. Carment et al. are concerned with at-risk states or "state fragility," which they admit is "an elusive concept." For them "fragile states lack the functional authority to provide basic security within their borders, the institutional capacity to provide basic social needs for their populations, and the political legitimacy to effectively represent their citizens at home and abroad." Lambach and Gamberger focus on predispositions to instability, and hope to contribute to a "unified concept" of political instability. Englehart and Simon present the most extensive discussion of the concept, its different dimensions, and possible different meanings and measurement.

Indeed, this lack of agreement on definition derives from a broader disagreement on the conceptual phenomena to be investigated (both in regard to intension, or the characteristics to be included, and extension, or the countries to be included in the definitional box). Some of this divergence in approach derives from a difference between basic and applied research. The former is focused primarily on explanations (or causes) for states that exist at some "fail<u>ed</u>" *end-point or condition*, while the latter is more concerned with a *process* in which at-risk, fragile, or fail<u>ing</u> states find themselves. For such policy-oriented analysts the problem

[1] While there are brief reviews of the literature in the other pieces, such as Iqbal and Starr, see Carment et al. for the fullest overview of current theory and findings.

is to understand the process in order to know how to intervene effectively to slow, halt, or reverse it. This is similar to the view of Englehart and Simon who also see two different views—state failure as a rare and extraordinary event (cataclysmic collapse) or a more widespread condition that should be investigated not as a discrete event, but rather as a condition that can be placed along a continuum of degree of failure, weakness, or risk.

If this characterization approximates what is actually going on, then there may be less confusion than appears at first blush. It may simply be a matter of scholars being much more clear and explicit as to research goals and the object of study – states we think have failed, or those which are failing. Two obvious analogies come to mind – the study of countries deemed *to be* democracies or the study of the process of *democratization*. Perhaps the most appropriate analogy is to a literature that studied the flip side of state failure – the study of social integration by which groups of people came together to create larger or more extensive political units! The study of international integration, based on the models of integration developed by Karl Deutsch and Ernst Haas (among others) was similarly beset by a confusion which rested on a debate over whether integration meant a process or an end-point. Depending on one's theory and research question, like integration, state failure can be either. It is up to each research project to clarify focus and purpose. It is hoped that the following studies will help the community of scholars to do just that.

References

Starr, H., 2006. *Approaches, levels, and methods of analysis in international politics: Crossing boundaries* (ed.). New York: Palgrave Macmillan.

———, 2006a. "Introduction: The future study of international relations, two-level games, and internal-external linkages", *Approaches, levels, and methods of analysis in international politics: Crossing boundaries,* H. Starr (ed.). New York: Palgrave Macmillan.

Werner, S., D. Davis, and B. Bueno De Mesquita, 2003. "Dissolving boundaries: Introduction", *International Studies Review,* 5:4, 1-7.

Pathways to State Failure

JACK A. GOLDSTONE

What is State Failure?

A great deal of attention has been paid of late to the problem of "state failure." Scholars are seeking to provide early warning for state failure, and policymakers are hard at work on devising ways to avert or remedy state failure (Esty et al., 1998; Rotberg, 2003; Krasner and Pascual, 2005).

But this begs the question—what is state failure? States can fail in a variety of ways, and from a variety of causes. Given this diversity of experiences, developing a precise and all-inclusive definition of state failure is a daunting task. What seems crucial is to find some middle ground between the view that there is a generalized set of causes that apply to all or most cases of state failure, and the opposite view that every state is different, and hence that every case of state failure is unique.

In this paper, I wish to suggest that we can identify two general qualities that states must possess to remain stable—*effectiveness* and *legitimacy.* Effectiveness reflects how well the state carries out state functions such as providing security, promoting economic growth, making law and policy, and delivering social services. Legitimacy reflects whether state actions are perceived by elites and the population as "just" or "reasonable" in terms of prevailing social norms. Thus, a state could be quite effective in raising revenues and maintaining security, but be seen as carrying out those tasks by excessive and arbitrary violence, and hence to lack legitimacy. Conversely, a state could be broadly inclusive and working within the law, and thus operating legitimately, yet be unable to agree on policies, to suppress violence, or to deliver economic growth or social services, and hence to lack effectiveness.

I shall argue that failed states are those that have lost *both* effectiveness and legitimacy. They neither function well, nor is their continued functioning desired by major groups of elites or the population. States that have high levels of *either* effectiveness *or* legitimacy, however, can survive for a number of years. A harsh dictatorship can survive for some years on effectiveness alone, or even for decades if it maintains high effectiveness and some degree of legitimacy. This is particularly the case if the state is effective in promoting economic growth and security, and relies on laws to some degree alongside coercion. Similarly, a popularly elected government that is seen to be pursuing just policies can survive for some years even if it has difficulty delivering on its programs, while it struggles to strengthen its capacities to govern.

Yet governments that lack either effectiveness or legitimacy remain unstable until they have acquired the other element. States depending only on effectiveness for support can fail when they lose the means to provide the economic benefits or coercive strength that sustains them. States depending only on legitimacy for support can fail when they reveal themselves to be corrupt, incompetent, too dependent on foreign powers, or become perceived as unjust through their policies or actions.

In sum, I shall argue that states with *both* effectiveness and legitimacy are usually resistant to failure; states with *either* effectiveness or legitimacy are unstable and prone to failure under changing circumstances; states with *neither* effectiveness or legitimacy are (or very soon will be) failed states.

These principles suggest a variety of ways in which a state can fail. A state that has high effectiveness but low legitimacy will fail if its effectiveness is compromised; a state that has high legitimacy but low effectiveness will fail if its legitimacy flags. States may also fail by suddenly shifting from having moderate levels of both effectiveness and legitimacy to having neither. Each of these theoretical patterns can arise from a variety of state actions or institutional changes.

I should also emphasize that while effectiveness and legitimacy may be independent conceptually, in the real world they are not wholly separate. Low effectiveness can undermine the legitimacy of a regime, as failure to provide security or economic growth or services will eventually be seen in itself as injustice. Similarly, low legitimacy can undermine effectiveness, as elites and popular groups will eventually cease to work for a regime that is seen as hopelessly unjust, undercutting the ability of the regime to function. Analysis of state failure thus needs to focus on the dynamics of how states build or lose effectiveness, how they acquire or forfeit legitimacy, and how these processes affect each other.

An Institutional Approach to State Failure

Scholars who study state failures—including revolutions, civil wars, and economic decay—have taken several different approaches to identifying the social conditions that lie behind these events. One approach focuses on long-term, deeply rooted characteristics of a society. These include the degree of trust or social capital developed over long periods (Putnam et al., 1994; Fukuyama, 1996; Coleman, 1990), the degree of economic inequality sustained over decades or centuries (Acemoglu et al., 2001; Engerman and Sokoloff, 2002), whether there are long-standing ethnic or religious fault-lines dividing a society (Horowitz, 1985), or structural conditions that lead to conflicts between the state and its own elites (Skocpol, 1979; Goldstone, 1991). Countries that have such deficiencies, it is argued, inevitably fail to make efficient use of their resources, suffer weak or declining economies, and fall behind other states. The combination of long-term decline and competition with other states leads to increased conflicts and state breakdown.

These long-term characteristics of societies may identify important and pernicious social patterns that produce state failure. Yet they leave policymakers with little choice except to triage societies and wait decades or centuries to see if such long-standing social characteristics will evolve and change.

A second approach lies behind the Failed States Index produced by the Fund for Peace and published by *Foreign Policy*. This approach seeks to identify a large number of discrete variables, each of which contributes to vulnerability to state failure. However, simply listing these factors provides no dynamic sense of how they combine, or with what weights, to lead to state failure. Thus the Failed States Index functions mainly as a checklist of items for policymakers to consider in appraising states' stability. It has only limited value as a guide to what *changes* or *processes* to look for or respond to in developing policies to strengthen or cope with states in the process of failing. The instability models of the Political Instability Task Force (Goldstone et al., 2005; Goldstone and Ulfelder, 2004) do estimate weights of different factors that make state failures more or less likely. Yet this predictive model again only identifies factors that make states vulnerable to crisis. The model provides no guidance on the dynamic processes that initiate state failures, nor any information on how the sequence of events in state failures unfolds.

A third approach has been taken by scholars who focus on the impact of institutions on current behavior. In the view of these "new institutionalist" scholars (Powell and Dimaggio, 1991; Brinton and Nee, 1998; North, 1990), people's behavior is shaped by the institutions under which they live. Good institutions produce good behavior and prosperous societies; bad institutions produce bad behavior and poor societies. Thus the new institutionalists argue that changing institutions can, in relatively short order, lead to changes in behavior—for good or for ill. This is because institutions provide incentives to individuals and groups to pursue certain activities; hence changing institutions creates new incentives, and hence different behaviors.

As one example, Yugoslavia held together as a peaceful multi-ethnic state under the communist institutions developed by Josip Tito for decades in the Cold War period. Tito's Communist Party allowed all ethnic groups in Yugoslavia a significant role in governance. Yet when Tito died, in the face of a massive economic crisis a new generation of nationalist politicians tried to reshape state institutions by emphasizing ethnic and religious identities and building political and military institutions based on those identities. The result was a major state failure, with civil war and genocide (Weingast and Figueiredo, 1999). Conversely, and more positively, a number of countries that were poor, ethnically or religiously divided, or had a recent past of political instability—and who are well outside regions that were historically favorable to political and economic development—have shown themselves capable of rapid and dramatic economic and political improvement. Examples include South Korea, Bangladesh, Uganda, and Malaysia. These cases argue strongly that *current* institutions can make a major difference.

Of course, even the new institutionalists realize that institutions are not easily changed. Grief (2006) has argued that institutions should be seen as self-reinforcing patterns of behavior, in which incentives generally encourage persistence and discourage change. To change institutions, it is necessary to substantially disrupt or alter these self-reinforcing patterns, so that new incentives, and hence new institutions, can take root.

Analyzing failing states, and finding ways to avert failure, thus depends on identifying clusters of institutions and incentives that produce stability, or that undermine stability when they change. The "tipping" points are not merely shifts in some index—such as employment, or income per capita, or deaths—rather, they are shifts in the perceptions and incentives embodied in institutional arrangements, such that people rather suddenly shift their behavior and allegiances to those institutions. We therefore need to analyze such institutions *as wholes*. When states are failing, lasting stabilization depends on rebuilding institutions in ways that provide lasting incentives for cooperative behavior.

In this paper, I shall present a number of institutional patterns of change that lead to state failure. The goal is not to be wholly comprehensive in enumerating every single pathway that could lead to state failure. The diversity of states and history implies that each failed state's pathway will have its unique elements. Rather, my aim is to map out several of the most common institutional patterns of change leading to state failure and their dynamics, and show how they can be described in terms of various combinations of loss of effectiveness and/or legitimacy. It is hoped this will better enable policymakers to recognize broad patterns of state failure when they begin to unfold, and aid in formulating responses that will grapple with these processes at the institutional level.

Pathways to State Failure: Dynamics of Institutional Change

I suggest there are five major pathways that comprise the most common processes leading to state failure. Each of these involves signals from the state to society that the effectiveness or legitimacy of the government is changing dramatically, or is about to change. These signals change the incentives of various groups to continue accepting government authority. These pathways are not exclusive, and may combine in various sequences.

1. Escalation of communal group (ethnic or religious) conflicts
   Examples: Rwanda, Liberia, Yugoslavia, Lebanon
2. State predation (corrupt or crony corralling of resources at the expense of other groups)
   Examples: Nicaragua, Philippines, Iran
3. Regional or guerrilla rebellion
   Examples: Colombia, Vietnam
4. Democratic collapse (leading to civil war or coup d'etat)
   Examples: Nigeria, Madagascar, Nepal
5. Succession or reform crisis in authoritarian states
   Examples: Indonesia under Suharto, Iran under the Shah, the Soviet Union under Gorbachev

Each of these stylized scenarios can be thought of as representing a pathway to state failure, or more analytically, a set of interactions by which the equilibrium underlying stable governance breaks down. In the real world, these are usually not encountered in "pure" form. For example, the collapse of the Soviet Union was mainly a reform crisis in response to Gorbachev's efforts to improve the popularity and efficiency of the Communist Party; but it also involved escalation of nationalist conflict over Russian control in the Baltic and Central Asian territories, and state predation (Communist Party corruption under Brezhnev) was also an issue.

In each of these scenarios, the state loses legitimacy or effectiveness or both, in regard to either the entire population or specific communal or regional groups. In the examples that follow, I illustrate how this occurs in each scenario. My emphasis here is on the *dynamics* of failing states. Thus I focus on common conditions that lead to changes in the effectiveness and legitimacy of state behavior as the best way to identify specific problems and possible remedies.

Escalation of Communal Group (Ethnic, Religious, or Other Identity Group) Conflicts

In an escalation of ethnic or religious conflicts, breakdown occurs because a major communal group or groups no longer have sufficient incentives to accept the government's authority, while the government is unable to prevent their rebellion. This pathway has two distinct modes.

The first mode occurs when a discriminatory but moderately effective government has excluded certain communal groups from power and the economy, but then loses effectiveness. The loss of effectiveness creates incentives even for former regime allies to seek a change in the government, while also encouraging excluded groups to seek redress or change.

A clear example is Liberia under the Samuel Doe regime. Prior to 1980, Liberia was dominated by Liberians who traced their descent from the American former slaves who had founded the country; native Africans faced considerable political and economic discrimination. Legitimacy was therefore weak, but the Americo-Liberians stayed in control by maintaining a moderately effective state. Then in 1980, President William Tolbert (a descendant of Americo-Liberians) was assassinated in 1980 by an army rebellion involving a small circle of enlisted men (all African-Liberians) led by Samuel Doe. Instead of opening his new regime to all Americo- and African-Liberians, Doe used his control of the government to enrich his own particular tribal group (the Krahn). Indeed, Doe seized control of Liberia's resources to enrich his tribal group more brutally and thoroughly than any prior Liberian regime, destroying the material security of all other groups, with Doe's army taking control of farms and businesses and overturning any previous checks on state and military authority. Doe's regime was also incompetent at governing, unable to provide either physical security or economic stability. Thus legitimacy and especially effectiveness *both* plummeted under the Doe regime. In 1989, in the wake of severe economic troubles, an opposition drawing on other tribal groups led by Charles Taylor and supported by other West African states invaded Liberia and initiated a civil war that devastated the country.

The second mode leading to escalating ethnic conflicts begins with an ineffective, but inclusive, democratic government. Despite being democratic, if certain regional or ethnic interests believe the regime is deeply corrupt, or tilting against their interests, they may decide to withdraw from and rebel against the government. If the government is too weak to suppress the rebellion, it can spread and threaten the regime. Democracy often arouses expectations of self-governance and greater rights in many ethnic groups in a multi-ethnic society. However, if the winning group in national elections in fact claims control of society and government, and acts in a corrupt manner or favors particular groups, the government may forfeit its legitimacy. This, plus the existing ineffectiveness, creates strong incentives for groups other than the winner to challenge the government or seek to change it.

An example is the failed democratic reforms in Nepal. In 1990, the monarchy of Nepal instituted a series of democratic reforms designed to create a multiparty parliamentary regime with the king as a constitutional monarch. However, feuding and corruption among party leaders, who used the parties mainly as vehicles for furthering their personal ambitions, and poor economic policies, cost the regime popular support. Rural people expected land reforms, which the new regime failed to deliver. In 1996, leaders of the Maoist Party in the Parliament withdrew from the government and moved to the countryside to pursue a guerrilla war against the government, calling for an end to the monarchy. The royal government was unable to defeat the insurgency, which at one point controlled nearly 70% of the countryside, and drew heavily on support from the Tarai and other rural ethnic groups who had been neglected by the Nepalese elite in Kathmandu. Nearly 10,000 people died in a decade-long civil war. In 2006, the monarchy agreed to terms with the rebels, who have established a new multi-party government, but the new regime is still being troubled by demands for greater autonomy and representation by rural ethnic groups.

To summarize, ethnic conflicts can lead to state failure under two distinct patterns of events: (1) a discriminatory regime that has been effective enough to prevent rebellion becomes ineffective at managing the economy or providing security, encouraging the groups that are being discriminated against to seek allies and rebel; or (2) a nondiscriminatory regime that has been inclusive but is weak or ineffective or starts to tilt toward a particular identity group and to exclude or damage others, thus encouraging groups to withdraw or seek to overthrow the regime.

State Predation

An authoritarian regime can stay in power a long time provided that it retains legitimacy in the eyes of crucial elites, such as the business community or the church or foreign supporters. State coercion is then backed by powerful economic and ideological support. However, once a regime is seen simply as predatory, enriching itself while posing a threat to all other elites and institutions, a broad antiregime coalition is likely to form that can mount an assault on the state.

An excellent example is the change in Nicaragua in the 1970s and 1980s. Nicaragua had long been ruled by a family dictatorship, the Somozas, established with U.S. support in the 1930s, and supported up through the 1960s by the business community and the church as well as the U.S. This dictatorship shared political power and economic wealth, to a degree, with the country's business elites in a shared-spoils system, while exploiting the peasantry. Although there were periodic peasant rebellions (such as that of the 1920s, which led to the rise of the Somozas), these were generally suppressed, and the Marxist opposition (the Sandinistas) remained small.

However, in the 1970s, the youngest member of the Somoza family took power; formerly head of the army, he had no use for business leaders and simply used his position to enrich himself. He used government revenues to purchase vast amounts of land for his family, and then in the mid-1970s committed two outrages. First, in the aftermath of the devastating Managua earthquake that destroyed much of the capital city, Somoza diverted aid revenues to his own pockets and failed to rebuild the city. Second, his forces assassinated a popular member of the business elite—Pedro Chamorro—whose media empire criticized Somoza's actions. This left Somoza with very low legitimacy among his key supporters, and he relied simply on U.S. backing and the National Guard's effectiveness to retain power. However, in the wake of these events, the business community and church establishment encouraged the U.S. to withdraw support from Somoza, and gave their support to the Sandinista insurgency. Thrown back solely on military support, Somoza waged a brutal war of repression against his enemies, but when he clearly lost U.S. backing, military desertions paved the way for Sandinista forces to take power.

Regional or Guerrilla Rebellion

The regional or guerrilla rebellion mode of state failure is similar to that of an ethnic group conflict, but a regional group or economic class, rather than an ethnic group, suffers discrimination or exclusion that is the locus of low legitimacy. Here, state failure arises when a group finds the government to be illegitimate, and then government ineffectiveness encourages other groups to join the rebel cause and renders the regime vulnerable to rebellion.

An example of a long-running regional rebellion is that of the FARC in Colombia. The rebellion began as a lower-class rebellion, but was sustained as the government security forces were not able to end the rebellion. Instead, rebel groups became entrenched in certain regions. Moreover, when the government of Columbia, with U.S. assistance, curtailed the Cali and Medellin cocaine cartels, the rebel groups (mainly the FARC) moved in to establish control of the cocaine growing regions. This gave the FARC the resources to expand its operations and challenge the government over a wider area.

Regional or guerrilla rebellions thrive on evidence that the state is weak, corrupt, or ineffective, and that the guerillas are better able to provide key services (including security) than the government. Violent strikes aimed at rebels, but which affect mainly civilians, often strengthen the rebellion and accelerate state failure by further depriving the regime of legitimacy while doing little to improve its effectiveness. Rebels succeed when the perception of the state as illegitimate and ineffective spreads, and the state's ineffectiveness in fact prevents it from suppressing the rebellion.

Democratic Collapse (into Civil War or by Coup d'Etat)

Democratic regimes usually have legitimacy to the extent that they include and respect the interests of diverse groups. The main problem of newly emerging democracies is usually ineffectiveness and inability to provide economic and physical security, although this is often exacerbated by loss of economic and political legitimacy where there is severe corruption. Democratic regimes can be paralyzed by factionalism, or rendered ineffective by lack of adequate resources to maintain the security of the population. In such conditions, democracies that are perceived as ineffective are often replaced by military regimes in coups. In many cases, lacking adequate control over its revenues, the government will become rife with corruption, which can further undermine its effectiveness as well as its legitimacy.

A typical example of democratic collapse occurred in Nigeria in 1983. The Second Nigerian Republic took office amid high hopes in 1979. However, the oil boom in Nigeria soon petered out, and seeking to gain support by pork spending, the government ran up debts of billions of dollars. Corruption flourished as well, and when the government managed to win reelection in a fraudulent election in 1983, further undermining its legitimacy, it was shortly overthrown by the army, which justified its actions both by the economic chaos and the rigged election.

Succession or Reform Crisis in Authoritarian States

The problems of succession or reform crises in authoritarian states arise directly from failure to manage the tradeoffs between state effectiveness and state legitimacy. In a succession crisis, a government whose legitimacy (or in some cases effectiveness) is dependent on the presence or political skill of a single powerful ruler finds itself in a quandary as to how to maintain power as that leader ages and approaches death, or if that leader dies unexpectedly. Unless a successor is groomed to be an effective ruler, and accepted as legitimate by the major business, military, and religious elites, the absence of such a successor leaves the state open to widespread potential opposition as different factions vie for power.

In most cases, if one faction aims to seize power, even if effective, they will be viewed as illegitimate. Regional or group-based rebellions are likely to arise. On the other hand, if some cooperative power-sharing arrangement is made to divide power, such arrangements may produce weak governance, leading to ineffectiveness and disintegration of the state as various regional or ethnic groups seek greater control over their own security. Succession crises thus often show up as a sudden shift of *both* effectiveness and legitimacy as the loss of the stabilizing personal leader grows closer or suddenly occurs.

The situation is well exemplified by Indonesia, where in the wake of an effort to replace the illegitimate succession of Suharto by a family member, the various power-sharing arrangements in parliament failed to produce a strong government and left various regions in turmoil as the army preyed on them and they sought their own security and autonomy. It took a series of elections and political party reorganizations, settlement of several regional independence struggles, plus UN intervention in East Timor to restore a measure of effectiveness and legitimacy to the Indonesian government.

A variation of this mode of state failure occurs when a state whose leadership wishes to increase its effectiveness or legitimacy embarks on reforms. In some cases, the reforms aimed at increasing effectiveness do little in that regard, but have the side effect of reducing legitimacy. For example, efforts to raise taxes, to add debts, or to manipulate the currency, as well as increases in corruption, may aim to raise the resources available to state leaders, but may not succeed in doing so. They may, however, weaken the state's perceived justice, and thus reduce its legitimacy. Conversely, efforts to raise legitimacy, for example by extending advisory roles to numerous groups, or having elections, may increase perceptions of justice. But they may also undercut the regime's ability to make and implement decisions, reducing effectiveness. Or, if the advice is ignored or the elections are rigged, such actions may have the reverse of the intended effect, and reduce rather than increase legitimacy. In short, there are many ways that leaders "get it wrong," and take actions that fail to deliver the intended results.

As one striking example of "reforms gone wrong," one can point to the breakup of Pakistan. In 1970, to increase the legitimacy of the Pakistani government, which was dominated by non-Bengalis from the western portion of Pakistan, general elections were held for a federal parliament. However, when Bengalis won a majority of seats, the non-Bengalis disavowed the results as unacceptable, leading to Bengali revolt, civil war, and the eventual partition of east and west Pakistan into Bangladesh and Pakistan, respectively.

A second example is the dissolution of the Soviet Union. President Gorbachev tried to increase the effectiveness of the Soviet regime by removing communist placeholders and corrupt party hacks. Unable to reform the party from within, he called for elections and external criticism to target and force out bad officials. Yet the calls for elections and criticism revealed such manifest incompetence and corruption in the party that the entire party lost any remaining legitimacy and became the target of calls for change. The party gained nothing in effectiveness, but lost all legitimacy; much elite, military, and popular support then swung to the anticommunist reformers led by Yeltsin.

In sum, there are at least five distinct patterns of institutional collapse by which states can approach state failure, and several sub-modes within those. All of these pathways involve some combination of the state's loss of effectiveness, or legitimacy, or both. The loss of effectiveness and/or legitimacy removes the incentives for elites and popular groups to accept the authority of the state, and creates opportunities and incentives for them to oppose it. States can survive, albeit in a fragile condition, if they are moderately effective, or moderately legitimate, even if the other characteristic of state capacity is weak. Yet when states become weak on both counts, total collapse is likely to follow.

Responses to Impending State Failure

One of the problems in preventing or responding to state failures is that such events have come at policymakers from many divergent angles. As we have just seen, state failures may arise from discriminatory regimes or inclusive democracies, from reforming autocracies or regional rebellions. However, there is a common thread in these diverse patterns, and that is the loss of effectiveness and/or legitimacy, especially in states that were already weak or lacking in one of these elements.

This argument would suggest that the way to prevent state failure is to identify states that are failing with regard to *either* legitimacy or effectiveness, but which have not lost both, and to intervene to shore up or restore the weak component of state capacity.

Yet this can be a tricky process, because of the tradeoffs that often occur between legitimacy and effectiveness. Making a regime more brutal and authoritarian can increase effectiveness, but at the cost of even greater loss of legitimacy. Launching democratic processes and holding elections can make a state more legitimate—but only if such elections are truly free and fair. Moreover, if the elections produce a stalemate, or results unacceptable to a major group or leader, then effectiveness of the state may be undermined. As Ottoway (1995, p. 239) points out, "Democratization is thus no panacea for a collapsing state. It can lead toward democracy, but it can also hasten state collapse." The very fact that state leaders themselves often hasten their own demise by trying to carry out reforms aimed at strengthening their position indicates how difficult is the task of averting collapse in failing states.

Responding to impending state failure is not as simple a matter as repairing a leaky roof on a house, where the technology is known and it is just a matter of applying well-understood remedies. It is more like trying to cure a cancer; for in the current state of knowledge, we do not know how to prevent or cure every case. Still, we can suggest some general guidelines for an overall approach.

A Determine whether the state in question is most endangered by low legitimacy, low effectiveness, or both. Which institutions or patterns of state behavior are causing a deficit in legitimacy—e.g., what is causing the state to be perceived as unfair or unjust? Which institutions or patterns of behavior are causing a deficit in effectiveness—is the state faltering in provision of security, economic growth, policymaking and enforcement, or social services?

B If a state has low legitimacy, but still has moderate to high effectiveness, one should determine whether it is likely that the government can be returned to legitimacy. Oftentimes, once trust has been lost, it cannot be recovered. If only one aspect of legitimacy is lacking (e.g., there is excessive corruption in government, or there is discrimination against a particular group, or the regime has been committing human rights violations), it may be possible to identify and correct that particular failing. However, if legitimacy is lacking on multiple dimensions, it may be better, if possible, to seek to support various actors that might have greater legitimacy, rather than to support a terminally illegitimate regime until it loses effectiveness and then undergoes collapse.

Leaders or institutions should be sought that have "preestablished" legitimacy (i.e., are considered fair and not tied to particular groups or interests) in the eyes of leaders and popular groups. These might be individuals who suffered under the old regime (e.g., Nelson Mandela in South Africa, Cory Aquino in the Philippines), or political processes that had legitimacy prior to the replaced regime (e.g., the use of a Loya Jirga council to choose leaders in Afghanistan). These situations—regimes with low or failing legitimacy—are likely to be the highest-cost and highest-risk interventions, but not impossible (e.g., 1986 Philippines).

Holding an election *may* be a way to increase the legitimacy of a regime, but it is far from a panacea. If an election is imposed on a highly illegitimate regime, most groups will consider the election itself likely to be rigged. An election that is not considered free and fair will not confer legitimacy, and indeed may trigger destabilizing protests. In addition, if various opposition groups regard each other with mistrust, elections may simply launch a new situation of instability, with factionalized groups creating a deadlocked, ineffective government or resorting to violence to reverse undesired electoral results. Replacing a corrupt and ineffective dictatorship with a similarly corrupt and ineffective democracy is *not* a recipe for avoiding state failure.

Elections are only likely to help avert state failure if several conditions are met: (i) that an electoral process be established that includes all key groups and provides for fair monitoring and supervision; (ii) that the major groups contesting the election, and their leaders, have a credible prior agreement to accept adverse results and honor guarantees to the losing parties regarding their rights, property, or other key concerns; and (iii) that institutions and practices are in place to limit corruption and provide security under the newly elected regime. Each of these conditions can be difficult to attain. If these conditions cannot be met, seeking an interim government under an alternative leader or council who has some prior legitimacy is likely a necessary precursor to elections. Otherwise elections are likely to accelerate, rather than avert, instability.

It is also important for outside actors to recognize that they cannot create legitimacy by "choosing" winners; indeed external support can often *reduce* the legitimacy of otherwise credible actors. For this reason, even when seeking to assist countries with rather illegitimate governments, it is important to act carefully and mainly in response to requests for assistance or support from within the society, rather than believing that it is possible to "bring in" legitimate actors or institutions from outside.

C If a state has high legitimacy, but low effectiveness (as with newly emerging democracies), it is more often possible to improve prospects for stability. In some cases, financial help will suffice, at other times organizational assistance. In many cases, third-party provision of neutral security forces to increase state effectiveness in safeguarding groups or property is sufficient to move toward stability.

However, most aid to legitimate but ineffective states has suffered from two drawbacks—(i) too much money is promised up front, more than the state's ineffective institutions can usually handle efficiently, and (ii) too little support is sustained over the longer term, when the state is able to deliver more and people expect more of the regime. Improving the effectiveness of a regime with moderate legitimacy is a long-term process, involving the construction of new institutions for administration, services, finance, and security. This is not something than can be accomplished by a short-term intervention.

D Actions to provide assistance to failing states need to be examined for their impact on *both* effectiveness and legitimacy. It is vital not to make a situation worse by policies meant to provide assistance but failing to attend to possible adverse consequences. For example, military or financial aid to an unjust regime may help keep its effectiveness up and prolong it in power, but it may also undermine its legitimacy, and thus do nothing to ameliorate, or may even worsen, its situation as an unstable state. A large flow of aid to an ineffective regime, if not carefully monitored, may create incentives for corruption that weaken the legitimacy of that regime. Similarly, elections designed to shore up legitimacy are likely to fail if they result in deadlock or infighting that undermines the effectiveness of the new regime. Assistance thus needs to be designed and implemented with attention to its effects on *both* legitimacy and effectiveness.

E Where many warning signs of failure are evident, the highest priority should be given to establishing *security and political legitimacy.* If these are established, other remedies to build effectiveness and legitimacy through economic growth and provision of social services can develop. However, if security is lacking, or if the political system is perceived as biased toward particular groups, other efforts to rebuild stability will likely be opposed by considerable elements in the population, and most often will be stymied or unproductive.

As two distressingly clear lessons in the potential value of this approach, we can look at U.S. policies in Afghanistan and Iraq. In Afghanistan, the U.S., after dispatching the Taliban, sensibly moved to find a way to build *legitimate* institutions. It utilized an extant institution with high legitimacy, the Loya Jirga, to develop a framework for government. Subsequent elections were reasonably fair and publicly supported, and the new democratic regime led by Hamid Karzai had substantial legitimacy. However, his regime was quite ineffective. It had no strong institutional framework for finance, security, administration, or service provision, and its ability to promote economic growth was weak. The U.S., however, was highly reluctant to extend the security assistance needed to help the Karzai government project its writ beyond Kabul, or to build strong central government institutions. Instead, U.S. policy accepted the regional power of local warlords, while promising more money up front than the Karzai government—given its weak institutions and limited reach—could effectively spend. Unfortunately, the lack of effectiveness of the Karzai regime had growing costs. The power of warlords, the growth of the drug trade as a key economic alternative, and the limited ability of the regime to project law enforcement and supply services undermined its legitimacy in the provinces and gave openings for the Taliban to regroup and mount a rebellion. If foreign powers had given equal attention to restoring the legitimacy *and* the effectiveness of the post-Taliban regime, some of the current difficulties in Afghanistan could perhaps have been reduced.

Conversely, in Iraq the U.S. poured troops and resources into Iraq for rebuilding. Yet, the procedures to create legitimate governance faltered. The figures originally backed by the U.S. (e.g., Ahmed Chelabi) proved to have weak support in Iraq, and the political processes the U.S. supported in building political parties and conducting elections worked to enhance divisions along regional and religious lines (Kurd, Sunni, Shia), rather than to bridge them and create a government with widespread legitimacy. The Shia-dominated government was not seen as legitimate, giving incentives for Sunnis (as well as foreign jihadist elements) to attack it. Such attacks eventually undermined the basic effectiveness of the government as well, overwhelming its ability to provide security, economic growth, or basic services. While the recent surge in U.S. troops and new tactics in cooperating with local recruits have increased the effectiveness of governance in Iraq, the question of legitimacy remains a problem. If U.S. policymakers had given more attention to the grave difficulties in developing a broadly legitimate government to replace the Baathist regime of Saddam Hussein, instead of focusing mainly on effective suppression of insurgents, the magnitude of current problems might have been diminished.

In sum, while the actual institutional pathways of state failure are diverse, the fundamental factors underlying state failures are loss of effectiveness and legitimacy. To avoid or respond to state failures, and to establish the foundations of strong and resilient states, state institutions with strong legitimacy and effectiveness must be developed. Keeping both objectives in mind throughout the design of policies to assist weak states is crucial to avoiding the problems that arise when states focus on legitimacy or effectiveness to the exclusion of the other. Both factors are equally critical to building strong states.

References

Acemoglu, D., S. Johnson, and J. Robinson. 2001. The colonial origins of comparative development: An empirical investigation. *American Economic Review* 91(5): 1369–1401.

Brinton, M. and V. Nee, eds. 1998. *The new institutionalism in sociology*. Stanford, CA: Stanford University Press.

Coleman, J. S. 1990. *Foundations of social theory*. Cambridge, MA: Harvard University Press.

Engerman, S. L. and K. L. Sokoloff. 2002. Institutions, factor endowments, and paths of development in the New World. *Journal of Economic Perspectives* 14(3): 217–232.

Esty, D., J. A. Goldstone, T. R. Gurr, B. Harff, P. T. Surko, A. N. Unger, and R. Chen, 1998. The state failure project: Early warning research for US foreign policy planning. In *Preventive measures: Building risk assessments and crisis early warning systems*, eds. J. L. Davies and T. R. Gurr. Boulder, CO: Rowman and Littlefield.

Fukuyama, F. 1996. *Trust: The social virtues and the creation of prosperity*. New York: Free Press.

Goldstone, J. A. 1991. *Revolution and rebellion in the early modern world*. Berkeley: University of California Press.

Goldstone, J. A., and J. Ulfelder. 2004. How to construct stable democracies. *Washington Quarterly* 28(1): 9–20.

Goldstone, J. A., J. Ulfelder, T. R. Gurr, M. G. Marshall, R. H. Bates, D. L. Epstein, M. Woodward, and M. Lustik. 2005. A global model for assessing political instability. Paper presented at the Annual Meeting of the American Political Science Association, Washington D.C.

Greif, A. 2006. *Institutions and the path to the modern economy*. Cambridge: Cambridge University Press.

Horowitz, D. 1985. *Ethnic groups in conflict*. Berkeley, CA: University of California Press.

Krasner, S. D. and C. Pascual. 2005. Addressing state failure. *Foreign Affairs* 84(4): 153–163.

North, D. C. 1990. *Institutions, institutional change, and economic performance*. Cambridge: Cambridge University Press.

Ottoway, M. 1995. Democratization in collapsed states. In *Collapsed states: The disintegration and restoration of legitimate authority*, ed. I. W. Zartman. Boulder, CO: Lynne Rienner.

Powell, W., and P. Dimaggio, eds. 1991. *The new institutionalism in organizational analysis*. Chicago: University of Chicago Press.

Putnam, R., with R. Leonardi and R. Y. Nanetti. 1994. *Making democracy work*. Princeton: Princeton University Press.

Rotberg, R., ed. 2003. *When states fail: Causes and consequences*. Princeton: Princeton University Press.

Skocpol, T. 1979. *States and social revolutions*. Cambridge: Cambridge University Press.

Weingast, B., and R. de Figueiredo. 1999. Rationality of fear: Political opportunism and ethnic conflict. In *Military intervention in civil wars*, ed. J. Snyder and B. Walter. New York: Columbia University Press.

The Logic of State Failure: Learning from Late-Century Africa

ROBERT H. BATES

Africa provides a unique and disturbing opportunity to explore the determinants of state failure. Angola, Chad, Namibia, Nigeria, and Sudan entered the 1970s war torn; by the mid-1970s, Sudan had exited from the list of civil wars but Equatorial Guinea and Zimbabwe had entered; by 1980, Zimbabwe had also exited, but been replaced by Mozambique, Nigeria, and Uganda. The pattern—a few dropping off the list of Africa's war-torn states, a larger number joining—continues into the early 1990s: only one country that was conflict ridden in 1990 became peaceful by 1992, while in the same period eleven new civil wars broke out. By the mid-1990s, Africa had become the most war-torn of continents. Why in late-century Africa did things fall apart? What can we learn from the politics of the continent about the failure of states?

To address these issues, I provide an informal characterization of a formal model that probes the foundations of political order. As the model has appeared elsewhere, I refrain from fully outlining its arguments and instead refer readers seeking further detail to the original publication (Bates et al., 2002). As we shall see, the model suggests that the continent was subject to a near-perfect political storm, the nature of which I examine in this article.

The Logic of Political Order

To address the collapse of political order in late-century Africa, I first turn to Weber (1958), who argued that the defining attribute of the state is its monopoly over means of coercion.

I depart from Weber in that I treat political order as a choice: the recent history of Africa reveals that heads of state can choose to behave as warlords (e.g., Robert Mugabe) or as statesmen (e.g., Nelson Mandela) and that citizens can choose to arm themselves (e.g., the Cobras in Brazzaville or the Lords Resistance Army in Uganda) or to live peacefully (as in Tanzania). *Order* prevails, I argue, when rulers choose to employ the means of coercion to protect the creation of wealth rather than to prey upon it and when private citizens choose to set weapons aside and to devote their time to the production of wealth and to the consumption of leisure. And a *state* is said to exist when the choices that characterize political order form an equilibrium.

By this line of reasoning, then, the conditions that make possible the state are those that underpin this equilibrium. To derive these conditions, we return to the model and provide an informal exposition of the logic underlying it.

The Motivation

Bates, Greif, and Singh (2002) base their argument on a scenario in which a community is peopled by a specialist in violence and two groups of citizens. Headed by powerful patrons, the groups can act in a unified manner.[1] The specialist in violence earns his living from the use of force; he either seizes the wealth of others or pockets funds that they pay for their protection. Sheltered behind powerful patrons, the citizens generate incomes by engaging in productive labor, but they too can be mobilized either to seize the income of others or to defend their incomes from seizure. The three personages repeatedly interact over time.[2] The question is: Can political order prevail in such a setting?

Strategies and Payoffs

The answer is: yes. Under certain well-specified circumstances, as Bates, Greif, and Sngh (2002) demonstrate, the specialist will choose to use his control of the means of violence to protect rather than to despoil private property. And the groups of citizens will choose to devote their time and energies to labor and leisure and forswear the use of arms, while rewarding the specialist in violence for protecting them against raids by others. Under certain well-specified conditions, these choices—which I label "orderly"—will persist in equilibrium, rendering political order a state.

Critical to the equilibrium is the prospect of triggering a punishment phase.[3] Whenever a citizen defects, as by raiding or failing to pay taxes, or whenever the specialist in violence defects, as by engaging in predation or failing to deter raids, then all the citizens take up arms and the specialist in violence turns from guardian to predator. The citizens re-arm so as to deter raiding by other citizens or by the specialist in violence. When doing so, they reallocate resources from production and leisure, with the result that their utility declines,

[1]That is, they have solved the collective action problem.

[2]Including by the specialist in violence, who can seize their assets only in probability and at a cost that is certain.

[3]As is well known, in a repeated game, multiple equilibria can exist; indeed, there can be infinitely many. In this article, however, it is not the game itself that is of interest; it is a particular equilibrium of that game—one that I have labeled "the state." The reason why governments and private agents assembled in ways that sustain this equilibrium lie outside of the scope of this paper; they have been addressed by those studying the legacies of imperialism, such as Coleman (1958). Young (1994); or the impact of the nationalist period, as in Hodgkin (1956).

The central concern of this paper is with the breakdown of that equilibrium.

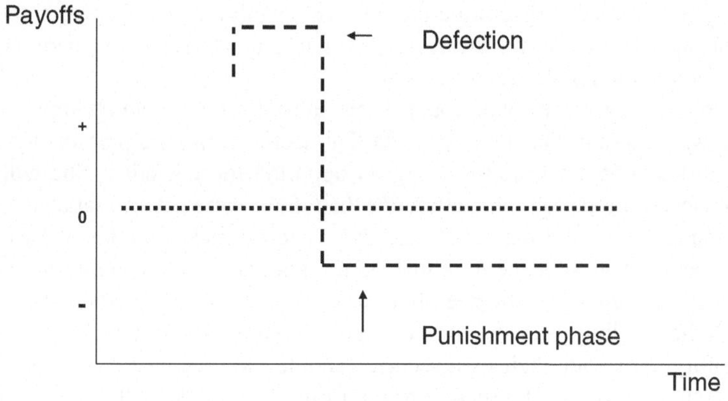

FIGURE 1 Payoffs on and off the equilibrium path.

as does the amount of income vulnerable to the warlord's predation. It is the prospect of triggering this reduction in payoffs that induces adherence to the use of orderly strategies.[4]

For a strategy to be in equilibrium, no player should be able to gain from deviating after any history. That is:

I. No agent should be able to gain by raiding or refusing to pay taxes.
II. Nor should an agent be able to gain by altering the allocation of her resources.
III. G's threat to predate—that is, the threat of the specialist in violence—must be credible.
IV. G must find it optimal not to predate if the private agents have refrained from raiding and paid their taxes.

The Conditions for the Equilibrium

To grasp the conditions under which this equilibrium is secured, focus on the choices open to the specialist in violence, as presented in Figure 1. In this figure, the vertical axis represents monetary gains or losses. The further above 0, the greater the payoffs; the further below, the greater the losses. The horizontal axis designates time, with the more immediate payoffs occurring near the origin and the more distant ones further to the right. The dotted line represents the flow of payoffs that result from tax payments; the flow is steady, moderate, and positive in value. The dashed line represents the flow of payoffs that result from predation. Predation yields an immediate benefit: the dashed line leaps above the dotted line, indicating that the income from predation significantly exceeds that from tax payments. But that one period spike then gives way to a stream of losses, as illustrated by the plunge below the zero point that separates gains from losses. Insofar as a decision maker is forward-looking, the losses that accrue in the punishment phase thus caste a shadow over the returns from defection and so should temper the wish to engage in predation.

If summed over time, each line—that representing the returns to taxation and returns to predation—yields an expected payoff. What would determine their magnitudes? In particular, what would determine whether the value of the variable path, generated by predation, will be more or less attractive than that of the steady path, generated from tax

[4]The analysis of a similar equilibrium in which the government punishes an agent who raided or failed to pay tax without reverting to the all defect equilibrium does not change the analysis.

payments? The factors that determine the relative magnitude of the payoffs determine as well the incentives to abide by the strategies that characterize political order and they therefore underpin the possibility of the state.

One is the specialist's rate of discount. A specialist in violence who is impatient, greedy, or insecure will discount the future payoffs that accrue along the path of play; she thus discounts the losses that accrue following an opportunistic deviation. She will therefore find the prospect of predation more attractive than if she were patient, satiated, or secure.[5]

A second is the level of taxes.[6] The tax rate is constrained from below by the need to induce the specialist in violence to withstand the temptation of engaging in predation. It is constrained from above by the need both to fulfill the citizens' participation constraint and to fulfill the credibility constraint of the specialist in violence. The first arises from the citizens' ability to provide their own security, should the specialist charge too high a fee for his protective services. The second arises from the possibility that the specialist will refrain from turning predatory should a citizen raid or fail to pay taxes, for fear of no longer being (abundantly) compensated by other tax payers. The equilibrium tax rate is thus knife edged.

The third condition is the magnitude of the rewards that predation might yield. If sufficiently lucrative, the specialist in violence might choose to deviate despite the costs of doing so, which would include the loss of tax revenues.[7]

Taken together, these conditions suggest that *the possibility of political order rests on the value of three variables: the specialist's rate of discount, the level of public revenues, and the rewards from plunder*. The interplay of these forces helps to determine whether governments safeguard or prey upon the wealth of the land, whether groups of citizens take up arms, and whether there is political order—or state failure.

This argument differs from other treatments of restrained predation, among the best known of which are those of Findlay (1991), McGuire and Olson (1998), or Wintrobe (1998). In these models, it is the *economic* response of private agents that constrain the specialist in violence:[8] the rational autocrat realizes that a moderate and steady rate of extraction will generate higher levels of revenues than will despoliation. In the model employed here, confiscation may in fact be profitable; the factor that stays the hand of the autocrat is the costs that subsequently arise from political disorder.[9]

In the works of Findlay (1991) and McGuire and Olson (1998), moreover, private agents take the actions of government as given. In doing so, they behave nonstrategically, responding to the government's choices by making income-maximizing choices of capital and labor. In this model (Bates, Greif, and Singh, 2002) as in Wintrobe (1998), private citizens behave strategically and by doing so shape the choices of their government. In Wintrobe's (1998) analysis, the private sector dissembles, falsifying its preferences (see also Kuran, 1995); in this, private agents control tax payments and possess arms, and threaten fiscal dearth and armed resistance, and—under the conditions set out above—dissuade the government from engaging in predation.[10]

[5]As is standard, I treat the discount rate as exogenously determined.

[6]Which is endogenously determined.

[7]In the arguments that follow, I treat the level of temptation as exogenous in the short run.

[8]Our phrase, not theirs.

[9]Phrased differently, Bates, Greif, and Singh (2002) shift the emphasis from the impact of taxation on the tradeoff between labor and leisure to the impact on participation and credibility constraints.

[10]Put another way, we depart from McGuire and Olson (1998) and Findlay (1991) by using sub-game perfection rather than Nash as our equilibrium concept.

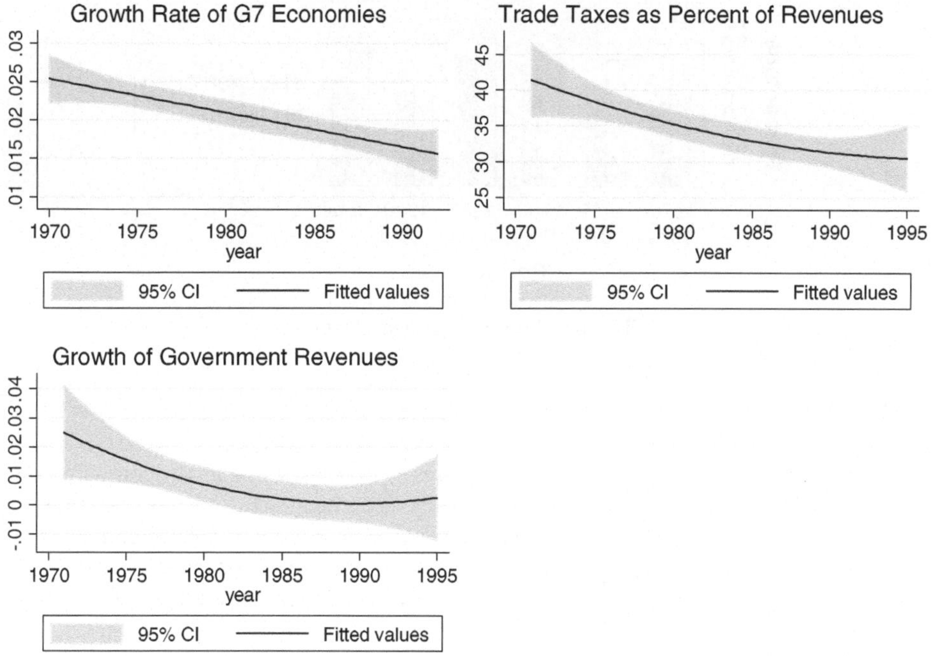

FIGURE 2 Revenue crisis of Africa's states.

Applying the Model

The recent political history of Africa is marked by the major economic and political shocks and the legacy of misguided policy choices made in the post-independence period (see Ndulu et al., 2007). The argument provides the causal mechanisms that link these forces to state failure. Each, it suggests, drove the value of a key variable out of the region in which orderly choices, by governments and private agents, constitute an equilibrium.

Fiscal Crisis

In the mid-1970s, political events in the Middle East drove up the price of oil in international markets and a cartel of oil-producing nations kept prices high until the mid-1980s. Because of higher energy costs, the advanced industrial economies entered a period of recession. For Africa, exports to the advanced industrial nations constituted a major source of income; governments, in particular, gained a major portion of their incomes from taxes on trade. As seen in Figure 2, the decline in the growth rate of the G7 nations found its parallel in reduced contribution of trade taxes to the governments' budgets in Africa—and to a decline in the size of the governments' budgets, measured as a percentage of the gross domestic product.

Adding impetus to this exogenous shock were forces unleashed by previous policy decisions. In the post-independence period, many governments in Africa had imposed interventionist economic policies. They had controlled the price of money, both limiting the rates of interest that banks could charge borrowers and the rate at which foreign currencies could be changed for local banknotes. They had regulated trade and controlled the prices at which manufactured goods could be sold, and most had regulated agricultural

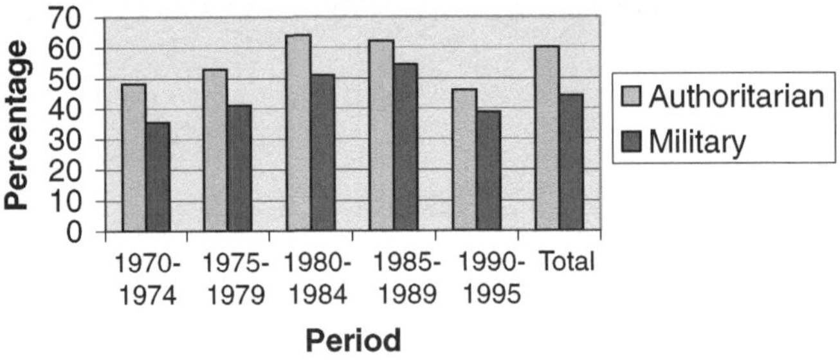

FIGURE 3 Political systems by time.

markets, compelling farmers to sell their produce at prices that lay well below that would have prevailed in competitive markets.[11]

The private economy reacted to such measures by fleeing the official economy. To avoid taxes on trade and the losses incurred at the official exchange rate, merchants smuggled goods across national frontiers. To avoid price controls, managers sold their product on the black market. To avoid costly regulation, entrepreneurs chose to invest outside of the formal economy. The expansion of the informal sector contributed to the shrinking of the tax base, as economic activity took shelter in the untaxed portions of the domestic economy.

In the later decades of the twentieth century, then, both exogenous and predetermined forces combined to shrink the fiscal basis of states in Africa.

Political Reform

According to the reasoning outlined above, the behavior of the government is shaped not only by the level of public revenues but also by the rate of discount. Should the specialist in violence become impatient, he would then have greater reason to defect, generating a sense of insecurity that induces members of civil society to forge their own means of defending life and property.

In late twentieth-century Africa, incumbent regimes were driven to revise their political expectations by forces unleashed at the global level, some economic and others political.

Economic Forces. As exports from Africa declined, public revenues fell and Africa's governments borrowed from abroad. By the mid-1980s, private bankers were no longer willing to lend and international financial institutions assumed the burden of Africa's debts. In order to increase the capacity of African economies to repay their loans, these institutions sought ways of enhancing their performance. In so doing, they increasingly focused on political reform, attributing the poor performance of the continent's economies to the authoritarian nature of its political regimes.

As creditors focused on the behavior of African governments, they struggled with the question: Why would these governments adopt policies that undermined economic prosperity? Over time, a consensus emerged: the behavior of these governments reflected their lack of political accountability. Not being accountable, governments in Africa could adopt policies that conferred concentrated benefits on the elites while imposing widely distributed costs on others. As can be seen in Figure 3, data drawn from 46 African states (see Table 2) in the 1980s indicate that in over 60% of the country-year observations, African governments were classified as authoritarian. In nearly one third, they were military; and if

[11]The best account is offered in Ndulu et al. (2007).

civilian, most were no- or single-party states, in which organized opposition and competition for office were banned. When pushing for greater accountability, international financial institutions sought the end of authoritarian forms of government.[12]

In their call for political reform, members of the international financial community were joined by those within Africa itself, and especially the better-off members of the urban sector who had witnessed the erosion of the quality of government institutions. On the one hand, this erosion took the form of reduced levels of service. On the other, it took the form of predation. Paid poorly, civil servants demanded bribes for services that the public had already purchased with their payment of taxes. And poorly paid soldiers began to prey upon private citizens, in some instances seizing the belongings of the citizens as if they were the possessions of an enemy (see, for example, Gould, 1980).

In response to the decline in the quality of public services and the predatory behavior of public employees, Africa's private citizens joined its international critics in calling for political reform, by which they tended to mean the reintroduction of multiparty systems and the selection of their rulers through open electoral competition (see Ake, 1996; Bratton and van de Walle, 1997).

Political Sources. Global recession and the legacy of past policy choices thus combined to create pressures for political reform. To a remarkable degree, Africa's authoritarian regimes were initially able to resist. A major reason, it would appear, were the services they rendered to those fighting the Cold War. Insofar as the political security of Africa's authoritarian regimes was underpinned by foreign patrons, they were unexpectedly rendered less secure when their patrons abandoned them following the collapse of communism.

Both Liberia and Ethiopia, for example, had permitted the United States to locate bases and to build air fields on their territories. Kenya had permitted the United States to upgrade the port facilities at Mombassa, thereby providing its navy a potential base in the Indian Ocean. And Zaire had leased air facilities to the United States and permitted its intelligence services to operate with relative freedom. While foreign critics and domestic reformers might condemn the public policies and political institutions that characterized these countries, their leaders—Master Sergeant Doe, Daniel Arap Moi, and Desire Mobutu, respectively—successfully sheltered themselves from the political consequences, counting on the backing of those who recruited local confederates in the struggles arising from the Cold War. Threats by international financial institutions to cut off lending to these governments were often overruled by security specialists more concerned with defeating communism than with global finances.[13]

As argued by Dunning (2004), with the end of the Cold War, economic threats now became credible. With the collapse of communism, those who presided over the military and political bureaucracies of the West no longer had reason to restrain those who staffed the international financial institutions. As Mobutu learned, for example, the Department of State would no longer check the demands of Treasury to pressure for the repayment of Zaire's debts and for the reform of its political institutions (Wrong, 2000). Should African governments seek economic concessions from those who held their debts, they now had to allow political opponents to organize, political parties to form, and political competitors to challenge them at the polls.

Economic and political challenges at the global level instilled fear among incumbents while giving hope to their opponents. As seen in Table 1, local political movements demanded the convening of constitutional assemblies, and some assumed sovereign status.

[12]See, for example, World Bank (1989; 1992).

[13]See Blumenthal, E. (1982) and Huband, M. (2001).

TABLE 1 The diffusion of political reform

			Election		Outcome: Incumbent	
Country	Date	Duration	Month	F&F?	Ousted	Retained
Benin	Feb-90	1 week	Feb-91	yes	√	
			Mar-96	yes	√	
Congo	Feb-91	3 months	Aug-92	yes	√	
Gabon	Mar-90	3 weeks	Dec-93	no		√
Mali	Jul-91	2 weeks	Apr-92	yes	√	
Niger	Jul-91	6 weeks	Feb-93	yes	√	
Burkina Faso	Aug-91	2 months	Dec-91	no		√
Ghana	Aug-91	7 months	Dec-92	yes		√
Togo	Aug-91	1 month	Aug-93	no		√
Zaire	Aug-91	1 year	–	–		
CAR	Oct-91	2 months	Aug-92	yes	√	
Chad	Jan-93	3 months	Jun-96	no		√

F&F: International monitors judge the election free and fair.

As the forces of reform gained power, incumbents were rendered insecure and their time horizons shortened. According to the logic of the argument, the costs of future penalties and the benefits of future tax payments would therefore both decline, increasing the temptation to deviate from the path of play that, when in equilibrium, yields political order.

Global economic shocks thus led to reduction in the value of public revenues, and global changes in the geopolitical system encouraged incumbents to discount their political futures more heavily. By the reasoning set out by Bates, Greif, and Singh (2002), both changes would have weakened the incentives for those who controlled the state to employ its power to safeguard, rather than to prey upon, the economy.

Natural Resources

A last element now comes into play: the economic environment within which these changes took place.

Many incumbents presided over economies richly endowed by nature, offering abundant rewards to those with the power to seize them. Even were elites to trigger state failure by engaging in predation, having appropriated rich resource rents, they could remain prosperous amid political disorder.

In Central and Southern Africa lie the mineral economies of Botswana, Zambia, Zimbabwe, South Africa, and Zaire. Along the western shores of the continent run the oil fields of Nigeria, Cameroon, Equatorial Guinea, and Angola. In Togo and Guinea there is bauxite; in Angola, Botswana, and Liberia, diamonds. In Ghana, South Africa, and Zaire, gold. And in eastern Africa, in the latter portions of the twentieth century, oil fields were being brought into production along the great rift.

The ruling elites could use their power to safeguard the production of wealth or to seize it. Were they to seize it, they would be regarded as predators and, by the logic of the model, private agents—firms and citizens—would then seek to provide their own protection, resulting in the militarization of political life and the spread of political disorder. But were incumbents to secure their income by tapping the rents that flowed from natural resources, they could prosper even in the midst of this disorder. They would therefore be more willing to deviate than would have otherwise been the case. And indeed, in Chad, Sudan, Nigeria, Angola, and elsewhere, Africa's political elites now course through national capitals filled with glorious buildings but populated by people maimed or orphaned by war (see Birmingham, 2002).

Discussion

According to the logic of the model, specialists in violence would refrain from predation and safeguard property rights so long as they were (moderately) well rewarded, secure, and faced future losses should they fail to do so. In late-century Africa, the conditions rendering restraint an equilibrium strategy abruptly altered. Exogenous contemporary shocks and the impact of past policy choices reduced the level of government revenues, while also increasing the pressures for political reform—and this in the midst of economies that promised alternative sources of income to those with the power to seize them. As the incentives to revert to predation rose, so too did the level of disorder. Societies began to militarize and states to fail in late-century Africa.

Others have of course stressed these three sources of disorder. Marxist scholars, such as Mlawa and Green (1995), have related economic collapse to political chaos in Africa; in this they have been joined by neoclassical economists, such as Collier and Hoeffler (2004), who find a statistically significant association between the level of poverty and the likelihood of civil war.[14] Political scientists, such as Mansfield and Snyder (1995), have addressed the impact of democratization upon civic conflict; in this, they have been joined by commentators, such as Zakaria (1997), who speaks of "illiberal democracies" and the ethnic and cultural violence that they foster. Lastly, Collier, and Hoeffler (2004) and a host of others (Cilliers, 2000; Klare, 2002; Ross, 2004) probe the impact of natural resources rents on political incentives, highlighting the political dimensions of the "resource curse."

It is one thing to assert such hypotheses; it is another to derive them as the implications of a theory (Hempel, 1965). The first is contention; the second, explanation. Moreover, it is one thing to defend the plausibility of an argument; it is another to test it. The first can be done by demonstrating that it is consistent with established facts, as has been done here; the second requires more systematic analysis. I therefore turn to the use of statistical methods.

Empirics

In doing so, I employ data from 46 African states, between 1970 and 1995 (Table 2).[15] I start with data from 1970 because the quantity of missing data increases with the depth of the panel. I end with 1995 because the major shocks had been processed by that time.

Dependent Variables

Throughout the following analysis, the appearance of reports of armed militias in standard references (see Table 3) serves as the dependent variable.

[14]See also Sambanis and Hegre (2006).
[15]The data are available from http://africa.gov.harvard.edu//.

TABLE 2 Countries in the sample

1. Angola	24. Madagascar
2. Benin	25. Malawi
3. Botswana	26. Mali
4. Burkina Faso	27. Mauritania
5. Burundi	28. Mauritius
6. Cameroon	29. Mozambique
7. Cape Verde	30. Namibia
8. Central Africa Republic	31. Niger
9. Chad	32. Nigeria
10. Comoros	33. Rwanda
11. Congo, Republic	34. Sao Tome and Principe
12. Cote d'Ivoire	35. Senegal
13. Djibouti	36. Seychelles
14. Equatorial Guinea	37. Sierra Leone
15. Ethiopia	38. Somalia
16. Gabon	39. Sudan
17. The Gambia	40. Swaziland
18. Ghana	41. Tanzania
19. Guinea	42. Togo
20. Guinea-Bissau	43. Uganda
21. Kenya	44. Dem. Rep. of Congo
22. Lesotho	45. Zambia
23. Liberia	46. Zimbabwe

I note the presence of militias because their formation signals a state's loss of the monopoly of violence. Particularly in Africa, body counts are inaccurate.[16] Mergers between and divisions among armed opponents of incumbent governments render the number of movements a misleading indicator of the magnitude of disaffection. And with challengers claiming hegemony over major regions and governments conceding the loss of small refuges, claims regarding territories controlled provide inaccurate measures of the magnitude of disorder. I use a qualitative measure, then, because I distrust the accuracy of the quantitative indicators available.

Independent Variables

As a measure of public revenues, I note the percent of the gross domestic product that accrues as revenues to the central government. The argument set out above highlights the importance of the level rather than of changes in public revenues on the likelihood of state failure. *I therefore expect the coefficient of the variable to be significant and negative in sign in pooled samples.*

The logic of the argument links changes to a competitive political regime to changes in leaders' discount rates. To test this argument, I employ information regarding the party system: whether it is no-, single-, or multiparty. The logic of the argument emphasizes the impact of changes rather than levels: it focuses on the impact of political reform. *I*

[16]See, for example, the appendix to Ellis (1999), as well as the critiques of body count data—something of great concern to the International War Crimes Tribunals—which are reported at www.microconflict.eu.

TABLE 3 Variables pertaining to political disorder

	Measure	Mean	Standard Error	Source
Dependent variable				
Militias	1 = Yes 0 = No	0.247	0.431	Data collected by research team from *Keesings Contemporary Archives, Africa Confidential, Economist Intelligence Unit*
Independent variables				
Revenues	Central government's revenues as percent of GDP	18.489	0.382	*World Development Indicators*
Petroleum	Value of exports per capita in constant U.S. dollars	87.010	14.331	Data collected by research team from commercial sources
Party system				
No-party system	1 = Yes 0 = No	0.355	0.015	Data collected by research team from *Keesings Contemporary Archives* *Africa Confidential* *Economist Intelligence Unit*
One-party system	ditto	0.445	0.015	
Multi-party system	ditto	0.205	0.013	
Duration				
No-party system	Length of time in years of duration of party system	2.405	4.685	Data collected by research team from *Keesings Contemporary Archives* *Africa Confidential* *Economist Intelligence Unit*
One-party system	ditto	3.415	5.551	ditto
Multi-party system	ditto	1.271	3.934	ditto
Neighbor average	Number of neighboring states reporting militias, civil, or international wars	1.514	1.953	Data collected by research team from *Keesings Contemporary Archives* *Africa Confidential* *Economist Intelligence Unit*

(Continued on next page)

TABLE 3 Variables pertaining to political disorder *(Continued)*

	Measure	Mean	Standard Error	Source
Privileged region	1 = yes 0 = no	0.891	0.331	Harvard research team
President from non-privileged region	1 = yes 0 = no	0.570	1.953	ditto
Period1	1 if 1970–74 0 otherwise			
Period2	1 if 1975–79 0 otherwise			
Period3	1 if 1980–84 0 otherwise			
Period4	1 if 1985–89 0 otherwise			
Time since last report	Count	5.993	6.235	

All independent variables lagged one year.

therefore expect the coefficients on the no- and single-party system to be negative in sign and statistically significant in fixed effects models, when the dummy variable indicating the presence of a multiparty system is dropped from the equation.[17]

In the reasoning set out above, the level of temptation of defection should be high in environments that contain abundant deposits of natural resources. *I therefore expect there to be a significant and positive relationship between measures of resource wealth and the likelihood of reports of militias.* As a measure of resource wealth, I make use of the per capita value of petroleum exports; having experimented with a wide range of alternatives,[18] I find this to be the most robust.[19] I also include an indicator that takes on the value "1" if the country contains an economically privileged region and a "0" if it does not. In the majority of the cases in which such a region exists, it is because it possesses favorable soils, produces export crops, or contains valuable deposits of petroleum, gem stones, or metals. As the variable Privileged Region is time invariant, I also introduce a variable that takes the value "0" when the president is from such a region and "1" when he is not; if having a resource rich region makes a difference, then this variable too should bear a significant relationship with the likelihood of reports of militias in the fixed effects model. *I expect the coefficient on Petroleum and Privileged Region to be statistically significant and positive in sign in both equations. I expect the coefficient on President from a Non-privileged Region to be statistically significant in both equations but have no expectations regarding the sign.*

The inclusion of the variables discussed thus far is motivated by our understanding of when political order will persist in equilibrium. In addition, I include a series of control variables, the inclusion of which is motivated by a desire to purge the estimates of several sources of error.

Missing Data. Cross-national data from Africa possess high levels of missingness (Honeker, 2000). Of most immediate significance is that when states fail, they are unable to produce national accounts. That the pattern of missingness bears a systematic relationship to the phenomenon being investigated poses a first-order challenge to testing.

In the jargon of statisticians, the pattern of missingness is MAR,[20] which poses both a danger and an opportunity. Should cases be dropped that contain missing values, the estimates would not only yield larger standard errors; they would also be biased. That the missing values bear a systematic relationship with ones that are observed offers a possible remedy, however, for the observed values can be employed to estimate the values that are missing (Schafer, 1997). Repeated runs of the imputation model yield a distribution of estimates of the missing values, with its own mean and standard error. Applying use of "Rubin's rules" (Rubin, 1996), standard models can then be applied to the multiple sets of imputed data to derive estimates of interest. To calculate these coefficients, I created five imputed data sets, each based on 1000 rounds of augmentation.[21]

I.I.D. While multiple imputation provides a means for addressing the most basic of the challenges, additional sources of error remain. One is the nonindependence of errors within panels; in response to this possibility, I employ robust standard errors, clustered by country. Nonindependence within the cross sections, as could arise from temporal shocks, poses a

[17]In models not reported here, I also note the impact of the shift from military to civilian regimes. The results are consistent with the results presented in Table 4, Equation 2.

[18]The value of diamond exports, mineral exports, resource rents, and so on. See Bates (2008).

[19]As does Fearon (2005).

[20]That is, missing at random, as opposed to missing completely at random.

[21]I note and stress that in imputing missing values I employ a large number of variables other than those that I enter into the estimation models. Their inclusion helps to correct for sources of missingness in addition to political conflict.

TABLE 4 Changes in the probability of militias

	Pooled Model		Conditional Logit	
	Coefficient	P > t	Coefficient	P > t
	(1)		(2)	
Revenues	−0.039 (−2.04)	0.061	−0.017 (−0.609)	0.556
Petroleum	0.000 (−0.328)	0.743	−0.001 (−0.349)	0.727
No-party system	−0.459 (−1.166)	0.246	−0.788 (−1.616)	0.107
One-party system	−1.017 (−2.004)	0.045	−2.166 (−4.071)	0.000
Duration				
No-party system	0.096 (2.789)	0.006	−0.0196 (−0.455)	0.649
One-party system	0.038 (1.179)	0.238	0.146 (3.642)	0.000
Multi-party system	−0.024 (−0.524)	0.600	0.065 (0.920)	0.357
Privileged region	1.403 (2.491)	0.014		
President from nonprivileged region	−0.567 (−1.679)	0.093	−0.807 (−2.027)	0.043
Neighbor average	0.092 (1.401)	0.161	−0.052 (−0.595)	0.552
Period1	−1.504 (−3.526)	0.000	−1.611 (−3.169)	0.002
Period2	−0.914 (2.615)	0.009	−1.398 (−3.565)	0.000
Period3	−0.039 (−0.148)	0.883	−0.076 (−0.228)	0.819
Period4	−0.183 (−0.667)	0.505	0.015 (0.051)	0.959
Time since last report	−0.188 (−5.089)	0.000	−0.017 (−0.68)	0.497
Constant	−0.385 (−0.584)	0.559		
Number observations	1048		813	

t-statistics in parentheses. In Equation 1, robust standard errors, clustered by country.

second; I therefore include Period Effects in all models. Because disorder in one country could affect the likelihood of disorder in another, I enter a variable Neighbor in all models (see Table 3), thereby correcting for this source of nonindependence as well. Lastly, there is the possibility that the likelihood of political disorder in a given year relates to the existence of political disorder in previous years. To correct for such time dependence, I include a variable, Time Since Last Report.

In addition to problems arising from the violation of the assumption of independence, applying standard regression procedures to a model that includes a qualitative dependent variable generates errors that are not identically distributed. I therefore employed maximum likelihood forms of estimation, using logit and conditional logit models.

Other Considerations. Experiments with the data suggested that the party systems—no-, single-, and multiparty—are not homogeneous: The impact of a party system depends upon the length of time that the system had been in place. To correct for unit heterogeneity, I therefore included a measure of Duration, or the number of years the party system had been in place.

Estimates of the relationship between the independent variables and the likelihood of state failure could also arise from the impact of omitted variables and reciprocal causation. In Equation 2 of Table 4, I introduce fixed effects in an effort to sweep out the impact of country-specific, time invariant characteristics, such as size, climate, geographical location, or colonial background. For reasons that are not fully understood, the inclusion of "privileged region" in the models frustrated attempts to locate valid instruments by which to correct for endogenity. I address the potential magnitude and direction of the resulting bias below.

Findings

As can be seen in Table 4, the coefficient on Revenues is significant in estimates derived from the pooled sample. Because of the impact of political disorder on the collection of public revenues, this estimate is biased downward. Specifications that lent themselves to the use of valid instruments suggest that the "true" value of the coefficient is in the neighborhood of −0.15, or four times greater than that reported in Table 4. Because the i.v. model that generated this coefficient does not take note of the presence or absence of a privileged region, I do not report it here.

The first measure of the magnitude of resource endowments, Petroleum, bears no relationship to the likelihood of state failure.[22] The second set of measures—the presence of a privileged region and the holding of office by a president from a nonprivileged area—do. While the first finding must be taken as evidence against the argument advanced in this paper, the coefficient on Privileged Region is consistent with it: the existence of abundant natural resources would appear to increase the likelihood that power will be employed in predatory fashion, inciting regional tensions, and thereby triggering the formation of militias. The interpretation of the coefficient on presidential origins is less obvious; its magnitude does, however, serve to underscore the significance of resource politics.[23]

Turning to the indicator of political risk, note the coefficients on the No-party and Single-party systems in Equation 2. Both provide a measure of the change in the likelihood of political disorder associated with the change to multiparty politics. Both are negative, suggesting that the introduction of open political competition was destabilizing. The standard errors suggest that the change was statistically significant when power had been held by a single ruling party.

The coefficients on the control variables indicate that both neighborhood and recency effects are weak; the latter is significant in the pooled sample, however. The coefficients on the period dummies suggest that the likelihood of state failure did indeed increase in the late 1980s. Changes in the variables included in the model capture a large portion of the observed relationship between period and state failure noted at the outset of this article, however.

The estimates remain robust to the inclusion of additional variables. These additional variables include the standard "modernization" variables: measures of education, income, and urbanization. They also include measures of shocks: economic shocks, such as terms of trade or growth shocks; climatic shocks, such as droughts; and political shocks, such as national elections. They include as well such standard controls as population, to capture the size of the country; and such nonstandard controls, as a measure of the extent to which ethnic groups sprawl across national borders.

Table 5 presents estimates of the magnitude of the coefficients based upon first differences, using the estimates derived from the pooled sample (King, 1998). In the case of Revenues, the data suggest that a shift from the level of revenues garnered in Sierra Leone

[22]Dropping "privileged region" and "president from a nonprivileged region" does not alter this finding. It therefore does not appear to result from multicolinearity.

[23]One possible interpretation is that, as argued by Azam and Mesnard (2003) and Acemoglu and Robinson (2001), placing power over wealth in the hands of the relatively poor increases the credibility of policies of economic redistribution, thus reducing the incentives to use military force to extract that wealth.

TABLE 5 Magnitude of effects

Variable	Pooled Model (First Differences) Percent Change in Probability	95% Confidence Interval	
Revenues From min to max 3.5 53.5	−0.314	−0.634	−0.089
Petroleum From min to max 0 6547.48	−0.121	−0.897	0.403
No-party system From 0 to 1	−0.070	−0.209	0.010
One-party system From 0 to 1	−0.111	−0.315	−0.007
Duration			
No-party system From min to max 0 25	0.486	0.171	0.734
One-party system From min to max 0 25	0.182	−0.056	0.548
Multi-party system From min to max 0 25	−0.054	−0.351	0.312
Privileged region From 0 to 1	0.231	−0.006	0.428
President from non-privileged region From 0 to 1	−0.072	−0.232	0.006
Neighbor average From 0 to 1	0.191	−0.047	0.518
Period1 From 0 to 1	−0.137	−0.353	−0.030
Period2 From 0 to 1	−0.103	−0.268	−0.017
Period3 From 0 to 1	0.005	−0.071	0.103
Period4 From 0 to 1	−0.023	−0.114	0.054
Time since last report From min to max 0 25	−0.382	−0.708	−0.127

in the mid 1980s to that in Botswana in the same period is associated with a 35% reduction in the likelihood of state failure. Insofar as political disorder depresses the level of public revenues, these estimates are biased downward and the coefficients conservative estimates of the political importance of public revenues. The data in Table 5 suggest that countries containing a Privileged Region are 23% more likely to experience state failure. Having a president from a nonprivileged region reduces the likelihood by 7 percentage points; the estimate is imprecise, however. Countries with no-party systems are 7% less likely to experience state failure than are those with multiparty systems and those with single-party systems 11% less likely; only the first estimate is significant, however, at $p < 0.05$. The estimates in Table 5 indicate that reports of militias were 10% to 14% less likely in the 1970s than in the 1990s. As seen in the coefficient to Neighbor Average, countries whose neighbors experienced more violence were more likely to experience disorder, although the coefficient is not statistically significant. The data in Table 5 suggest that a country without unruly neighbors, such as Mauritius, is roughly 20% less likely to experience political disorder than, say, Zaire, which in 1994 bordered on six war-torn neighbors.

Conclusion

This article employs the logic of the model of Bates, Greif, and Singh (2002) and materials from late-century Africa to probe the foundations of political order. Bates, Greif, and Singh (2002) specify the conditions under which states become possible. By the same token, their argument specifies the conditions under which "orderly strategies" no longer can be sustained in equilibrium and states therefore fail. The materials from late-century Africa highlighted the plausibility of this argument while the model provided the causal linkages linking major episodes of Africa's recent history—the debt crisis, the end of the Cold War, and the launching of political reforms, for example—with the rise of political disorder and the collapse of Africa's states.

The mark of state failure is the government's loss of its monopoly over the means of coercion. Statistical estimates demonstrated that the likelihood of reports of militias correlated with the level of public revenues, changes in the level of political insecurity, and measures of resource wealth. While the methods employed cannot support causal inference, had correlation *not* been observed, the power of these methods would have been sufficient to reject causal claims. Bringing the logic of Bates, Greif, and Singh (2002) to bear on materials from late-century Africa has thus deepened our insight into the nature of political order and the underpinnings of the state.

References

Acemoglu, D., and J. A. Robinson. 2001. A theory of political transitions. *American Economic Review* 91: 938–963.

Ake, C. 1996. *Democracy and development in Africa*. Washington, DC: The Brookings Institute.

Apter, D. 1964. *Ghana in transition*. New York: Atheneum.

Azam, J.-P., and A. Mesnard. 2003. Civil war and the social contract. *Public Choice* 115: 455–475.

Bates, R. H. 2008. *When things fell apart: State failure in late-century Africa*. New York: Cambridge University Press.

Bates, R. H., A. Greif, and S. Singh. 2002. Organizing violence. *Journal of Conflict Resolution* 46(5): 599–628.

Birmingham, D. 2002. Angola. In *A history of postcolonial Lusophone Africa*, ed. P. Chabal. London: Hurst & Company, pp. 138–184.

Blumenthal, E. 1982. Zaire: Rapport sur sa credibilite financiere internationale. *La Revue Nouvelle* 77(11 November): 360–378.

Bratton, M., and N. van de Walle. 1997. *Democratic experiments in Africa*. Cambridge: Cambridge University Press.
Cilliers, J. 2000. Resource wars: A new type of insurgency. In *Angola's war economy*, ed. J. C. a. C. Dietrich. Pretoria: Institute for Security Studies.
Coleman, J. S. 1958. *Nigeria: Background to nationalism*. Berkeley and Los Angeles: University of California Press.
Collier, P., and A. Hoeffler. 2004. Greed and grievance in civil wars. *Oxford Economic Papers* New Series 56(4): 563–595.
Dunning, T. 2004. Conditioning the effects of aid: Cold War politics, donor credibility, and democracy in Africa. *International Organization* 58(2): 409–423.
Ellis, S. 1999. *The making of anarchy*. New York: New York University Press.
Fearon, J. D. 2005. Primary commodities exports and civil war. *Journal of Conflict Resolution* 49(4): 483–507.
Findlay, R. 1991. The new political economy: Its explanatory power. In *Politics and policy making in developing countries*, ed. G. M. Meier. San Francisco: International Center for Economic Growth.
Gould, D. 1980. *Bureaucratic corruption and underdevelopment in the Third World: The case of Zaire*. London: Pergamon Press
Hempel, K. 1965. *Aspects of scientific explanation*. New York: Free Press.
Hodgkin, T. 1956. *Nationalism in colonial Africa*. New York: New York University Press.
Honeker, J. 2000. *Issues in multiple imputation of data of the African Research Program*. Cambridge, MA: Department of Government, Harvard University.
Huband, M. 2001. *The skull beneath the skin: Africa after the Cold War*. Boulder, CO: Westview.
King, G. 1998. *Unifying political methodology: The likehood theory of statistical inference*. Ann Arbor: University of Michigan Press.
Klare, M. T. 2002. *Resource wars*. New York: Henry Holt and Company.
Kuran, T. 1995. *Private truths, public lies*. Cambridge, MA: Harvard University Press.
Mansfield, E., and J. Snyder. 1995. Democratization and war. *Foreign Affairs* 74(3): 79–97.
McGuire, M. C., and M. Olson. 1998. The economics of autocracy and majority rule. In *The political dimension of economic growth*, eds. S. Borner and M. Paldam. New York: St. Martin's Press: 38–73.
Mlawa, H. M., and R. H. Green. 1995. *Through structural adjustment to transformation in sub-Saharan and Africa*. Dar es Salaam: Dar es Salaam University Press.
Ndulu, B., P. Collier, R. H. Bates, and S. O'Connell. 2007. *The political economy of economic growth in Africa, 1960–2000, 2 vols*. Cambridge: Cambridge University Press.
Ross, M. L. 2004. How does natural resource wealth influence civil wars? Evidence from thirteen cases. *International Organization* (Winter).
Rubin, D. B. 1996. Multiple imputation after 18+ years (with discussion). *Journal of the American Statistical Association* 91: 473–489.
Sambanis, N., and H. Hegre. 2006. Sensitivity analysis of empirical results on civil war onset. *Journal of Conflict Resolution* 50(4): 508–535.
Schafer, J. L. 1997. *Imputation of missing covariates in multivariate linear mixed model*. University Park, PA: Department of Statistics, The Pennsylvania State University.
Weber, M. 1958. Politics as a vocation. In Max Weber, eds. H. H. Gerth and C. W. Mills. New York: Oxford University Press
Wintrobe, R. 1998. *The political economy of dictatorship*. Cambridge and New York: Cambridge University Press.
World Bank. World 1989. Sub-Saharan Africa: From crisis to sustainable growth. Available from World Bank Archives, http://web.worldbank.org.
World Bank. 1992. *Governance and Development*. Washington, DC: The World Bank.
Wrong, M. 2000. *In the footsteps of Mr. Kurtz*. London: Fourth Estate.
Young, C. 1994. *The African colonial state in comparative perspective*. New Haven and London: Yale University Press.
Zakaria, F. 1997. The rise of illiberal democracy. *Foreign Affairs* 76: 22–43.

Bad Neighbors: Failed States and Their Consequences

ZARYAB IQBAL
HARVEY STARR

Introduction

The issue of "failed" states, including the negative consequences of state failure, has recently drawn significant attention from a number of international actors. Failed states are seen as being linked to a range of problems—economic, social, political, and military. And they are associated with a wide range of negative consequences for their own populations, their neighbors, their regions, and the global community. *The Economist* (2005, 45) stated that "[t]he chief reason why the world should worry about state failure is that it is contagious." A similar concern has been echoed by international organizations such as the World Bank, by state governments—including those of the United Kingdom and the United States (see for example DFID, 2005; Goldstone et al., 2000)—and private foundations, such as the Fund for Peace of the Carnegie Endowment for International Peace (Fund for Peace, 2006). The concern within academic as well as policymaking circles about the collapse of weak states is to a great extent due to the possible negative impact of such occurrences at the international level. In this study, we examine the likelihood of regional diffusion of state collapse and the key negative consequences of state failure, including political instability and civil and interstate war. Specifically, we argue that failure in a state leads to destabilizing effects in neighboring and nearby states.

State failure may result in political disturbances ranging from minor political unrest to interstate war. Of concern here is the question whether these effects of state failure spread to neighboring states, and whether states in close proximity of a failed state have a higher likelihood of experiencing a collapse. The consequences of state collapse we study include political unrest and instability, civil war, international conflict, and state failure itself. Studies over the past forty years have shown that the causes of violent conflict—both external and domestic—include previous violent conflict. Conflict has been found to act as both cause and consequence of further conflict. Similarly, we consider state failure, and its effects, as both the consequence and cause of further state failure. Evidence of whether and to what extent state failure and its effects diffuse is central to policy questions on ways to address state failure.

First, however, we provide a brief description of our operationalization of state failure, which we believe to be a superior measure to existing conceptualizations of state collapse. Following the discussion of our definition of state failure, we present our expectations regarding the diffusion of state failure and its consequences. Next, we outline the data and operationalization used in the empirical analysis and present a discussion of the results. The final section concludes with a brief comment on the implications of our findings for future work on the subject of state collapse.

Measuring State Failure

A variety of definitions and a range of indicators of state "failure" or "fragility" exist in various reports and research endeavors. Here we use the conceptualization of state failure developed in a preliminary study of the concept and causes of state collapse by Iqbal and Starr (2007). As discussed in that work, earlier identification exercises have been quite incomplete in their conceptualization of "failed states." Not only have these treatments been incomplete, however, but they are highly problematic in that they have been essentially circular in their linking of concepts and measures, thereby creating considerable difficulties in research design. Specifically, most existing measures or indices of state failure incorporate a number of factors that may, in fact, be determinants of state collapse—such as civil strife and poverty. This causes endogeneity issues that make it a challenge to study state failure as a dependent variable.

A number of organizations have contributed to existing definitions of state failure, including the U.K. Department for International Development (DFID) (2005), The World Bank (2002), the Fund for Peace (2006), and the State Failure Task Force (Goldstone et al., 2000). For instance, the DFID outlines four broad categories of "indicative features of fragile states": state authority for safety and security, effective political power, economic management, and administrative capacity to deliver services; each is categorized in terms of the "capacity" and the "willingness" to provide them. The World Bank (2002) describes the LICUS (Low-Income Countries Under Stress) countries as "fragile" countries that are "characterized by a debilitating combination of weak governance, policies and institutions, indicated by ranking among the lowest (<3.0) on the Country Policies and Institutional Performance Assessment (CPIA)." The World Bank states that a "definitive list of LICUS is impossible to draw up since LICUS-defining characteristics are a continuum ... However, there is a consensus for analytic and operational purposes that some countries' policies, institutions, and governance can be defined as exceptionally weak when judged against the criterion of poverty reduction, especially with respect to the management of economic policy, delivery of social services, and efficacy of government." The World Bank does not disclose a complete list of the indicators used in the CPIA or how the score was produced.

Another definition of state failure, offered by the Fund for Peace (2006), states that a "state is failing when its government is losing physical control of its territory or lacks a monopoly on the legitimate use of force. Other symptoms of state failure include the erosion of authority to make collective decisions, and inability to provide reasonable public services, and the loss of capacity to interact in formal relations with other states as a full member of the international community." The Fund for Peace (2006) uses a list of 12 indicators to evaluate state stability; these indicators include demographic pressures, refugees and displaced persons, group grievance, human flight, uneven development, economic decline, delegitimization of state, public services, human rights, security apparatus, factionalized elites, and external intervention. In yet another conceptualization exercise, the State Failure Task Force collected data on 1231 variables for 195 distinct countries between 1955 and 1998 (King and Zeng, 2001, 655). According to the Task Force report (Esty et al., 1998, 27–38), "state failure" included: revolutionary wars, genocides and politicides, and adverse or disruptive regime transitions.

Some of these definitions focus on the loss of governmental control, while others are more concerned with socioeconomic conditions and poverty. Although addressing important issues, they fail to provide us with a definitive and measurable conceptualization of state failure. Moreover, these definitions suffer from a fundamental conceptual flaw, in that they all use the results of the analyses (which are based on the independent variables) to describe the concept being studied—that is, the dependent variable. To avoid these problems of endogeneity in the operationalization of state failure, we extract a measure of state failure from the POLITY IV data (Marshall and Jaggers, 2004) that is independent of key determinants of state collapse, such as armed conflict and poverty. POLITY IV offers three "standardized authority" codes that denote periods during which the regular government of a state has been interrupted and "state authority patterns are volatile or not politically salient" (Marshall and Jaggers, 2004, 17). The authority code of "Interruption Period" in POLITY (coded −66) indicates when a country is occupied by a foreign power during armed conflict. A "Transition Period" (coded −88), reflects a period before a regime change, particularly for democratic or "quasi-democratic" areas that are moving to full independence.

We use the category referred to as "Interregnum Periods" (coded −77) in this project as a measure of state failure. According to the POLITY IV *Dataset Users' Manual* (17–18), "A '−77' code for the POLITY component variables indicates periods of 'interregnum,' during which there is a complete collapse of central political authority ... Interregnal periods are equated with the collapse, or failure, of central state authority, whether or not that failure is followed by a radical transformation, or revolution, in the mode of governance" (Marshall and Jaggers, 2004). These periods are distinct from periods of instability and/or war during which a state's governing institutions remain in place, and, in fact, reflect times during which the main governmental authority has experienced a complete collapse. During the time period considered in our analysis, Lebanon was coded as a failed state 16 times and Laos failed 12 times; these states experienced a complete collapse of central authority during the highest number of years in the data. On the other hand, some states with protracted civil wars (such as the Sudan) did not experience a failure due to perpetuation of a central government.

The observations coded −77 in the POLITY data are not attached to any specific theory or model and, most importantly, are not based on a list of factors that are used as both independent variables and to measure the dependent variable. The −77 category, therefore, provides us with an unambiguous and explicit measure of state failure, which lends itself very well to empirical analysis. Thus, using POLITY IV, we have a set of completely "collapsed" states from 1946 to 1999, which allows us to avoid the conceptual/circularity problems inherent in extant definitions of state failure. With this empirically viable measure of state failure in hand, we can turn to assessment of the prospects for diffusion and contagion effects in state collapse.

Diffusion/Contagion and State Failure

In general, the analysis of state failure can be seen as an exercise in policy evaluation. Previous investigators, as noted, have considered this a dangerous phenomenon, with dire consequences. Working from this belief, policy prescriptions have been made in order to forestall such negative consequences. However, before we can have much confidence in such policy prescriptions, a number of analytic tasks must be undertaken. Previous work (Iqbal and Starr, 2007) has identified a set of key factors or conditions that increase the probability that a state will "fail." These factors include political instability, domestic unrest, civil war, and international conflict. That analysis revealed that both civil war and international conflict increase the likelihood of state failure, with civil war displaying a significantly stronger impact. Measures of lower levels of political upheaval at the domestic level—political instability and unrest—were also positively associated with state collapse. A central question here is whether the existence of the factors associated with causing state failure—instability, unrest, civil war, and international conflict, as well as state failure itself—leads to similar conditions in neighboring and nearby states. That is, does state failure have the negative consequences that more anecdotal/less rigorous analyses have suggested? If so, then concerned investigators could move on to the policy prescription and evaluation stage.

The negative consequences of state failure noted above would result from some sort of diffusion processes. Most and Starr (1980) open their article on diffusion, reinforcement, and the spread of conflict by noting that the "notion that an event may alter the probability of subsequent events through diffusion or contagion processes is not new" (932). While three decades ago there were relatively few researchers studying diffusion, the application of spatial diffusion or contagion models has become far more common, even if only implicitly referenced through the use of contiguity as an independent, intervening, or control variable. The interest in the impact of geography and location of states on various issues in international relations continues to grow (e.g., Beck et al., 2006; Gleditsch and Ward, 2002; Simmons and Elkins, 2004). The diffusion approach itself reflects an important theoretical position. Diffusion studies are valuable in their recognition that events external to some social unit have consequences for or effects on those units; affecting the probability that similar events will or will not occur. One major concern of Most and Starr (1980) was to highlight the proposition that large-scale conflict events were not independent, and, thus, that there were significant effects on standard research design and the standard use of statistics. O'Loughlin and Witmer (2005, 10), in a recent review of approaches to the study of civil war, reiterate this point: "Most analyses of social science data have proceeded apace with an implicit assumption that all the data are generated by a random process that results in the data being independently, identically, distributed." Diffusion analyses directly address this core assumption.

Some diffusion approaches also recognize the spatial element of the consequences or effects of events/actions—going beyond simple emulation or prototype-cueing processes—to argue that proximity to some stimuli increases (or decreases) the probability of subsequent behavior or events in nearby social units (see Starr, 2003; also Most and Starr, 1990). The term "contagion" reflects an epidemiological perspective that sees certain events or stimuli spreading through various forms of contact. Positive spatial diffusion through close proximity (such as contiguity) could be conceptualized as "skin-borne" disease. Regional effects could perhaps be seen as "air-borne" disease—where proximity is still important, but less so than in contiguous situations. It is crucial to distinguish these "contagion" processes from the conceptualization of diffusion simply as emulation, whereby people see some occurrence—no matter how near or far—and change the probabilities of their own behavior through a desire to repeat (or avoid) that occurrence.

The negative view of state failure held by international organizations and state governments rests on perceptions—or assumptions—that there will be a positive spatial contagion of the negative aspects/effects of failure through the spread of the disease to contiguous neighbors or the near region. We are concerned with whether the instability, unrest, civil war, or interstate war connected to state failure/collapse exhibit such spatial spread. Echoing Most and Starr's (1980) analysis of the diffusion of large-scale social conflict (both internal and external conflict), there are important theoretical reasons why one could hypothesize positive spatial diffusion. As has been argued (e.g., Most and Starr, 1980; Siverson and Starr, 1991), large-scale conflict such as civil war or interstate war, or highly destabilizing situations as state collapse, can significantly affect the opportunities and willingness of nearby areas. Such activities may affect the expected utility of attacking weak nearby targets, or becoming a target oneself from the mobilized armed forces of neighbors (e.g., Starr, 1994). State failure, like civil war, might make states inviting targets for other states; yet any violent and/or large-scale conflict could promote cascading internal effects in any of the participants—making additional subsequent state failure (or its effects) more likely.

Drawing on the work of Midlarsky (1975), nearby conflict (or here, state failure) may generate various types of uncertainty and anxiety, and could affect the probability of certain types of conflictual behavior. Looking at the occurrence of conflict and the possible effects on neighbors, Most and Starr (1980, 935) note that the "important point ... is that each nation's structure of risks and opportunities is likely to be changed once a war is under way and these changes may be most dramatic for those nations which are proximate to the warring nations."

Here we make a similar argument about state collapse and its effects on neighbors and neighboring regions. These arguments are also reflected in the following observations by O'Loughlin and Witmer (2005, 3) regarding the influence of geographic context:

> A lack of territorial sovereignty and an inability to form a national identity conspire to keep weak states vulnerable to volatile domestic circumstances. The spatial clustering of weakened states, and the subsequent clustering of conflict in weak states, allows for conflict to cross borders, infecting already vulnerable states. Therefore, the location of a state (and its civil wars) *is not simply an attribute, but another potential cause of conflict* [emphasis added]. States with high risk are subject to increased risk because 1) neighboring wars exacerbate volatile domestic conditions inside bordering states, and 2) neighboring wars can (and frequently do) spread into nearby states. Weak states cannot mitigate conflict diffusion and escalation from outside state borders.

While a diffusion approach focuses on external factors and effects, it must always be recognized that the unit of analysis of interest must have some combination of factors that make it "ready for diffusion," thus pulling together both external and internal factors. Here, we hypothesize that negative causal elements for state failure—instability, unrest, civil war, and external war—will diffuse from failed states to nearby states. Given the existing literature demonstrating general diffusion processes at work for both interstate and civil war (see, for example, Most and Starr, 1980; Starr and Most, 1983, 1985; Siverson and Starr, 1991; Gleditsch and Ward, 2000; Gleditsch, 2002; Sambanis, 2002; or O'Loughlin and Witmer, 2005; Salehyan and Gleditsch, 2006), we think this is a reasonable hypothesis. The diffusion of these conflict-oriented factors will, in turn, make nearby states "ready" for the diffusion of state failure. Thus, we will address a second hypothesis, that another negative consequence of state failure—failure itself—will also be found to diffuse through spatial contagion.

It should be clear, however, that we are not attempting to develop a full model of inter- or intrastate conflict, nor are we attempting to develop a full model of the causes of state failure. The exact conditions which make any state "ready" for "failure" may vary considerably as the literature noted above indicates. The mechanisms by which failure or its effects spread, are also varied—the diffusion perspective is based on the notion that proximity generates multiple modes of transactions, including flows of people, trade, information, military actions, and so forth. Flows of refugees, ethnic groups that cross borders, the existence of rebel sanctuaries, intergovernmental military cooperation or conflict, economic sanctions or assistance, and the like, will all affect opportunity and willingness.[1] With the many variables and possible combinations of variables, there is a degree of freedom problem—many variables/combinations and only a limited number of state failures. Future work employing comparative case studies would help to uncover which mechanisms tend to be at work under different conditions.

Data and Operationalization

We argue that in addition to devastation in the state that experiences it, state failure can result in the spread of destabilizing elements at the regional level. Specifically, we examine the likelihood that the political consequences of state failure would spread to other states; these consequences are political unrest and instability, civil wars, subsequent state failures, and international war. Our empirical analyses, therefore, are designed to evaluate whether—and to what extent—the negative consequences of state collapse diffuse to neighboring and nearby states. We examine regional diffusion of the key consequences of state failure for the period 1946–1999, with country-years as the unit of analysis.

Dependent Variables

Given our interest in the destabilizing effects of state failure, we focus on five negative consequences that might spread to other states in the region. The first two of these deal with lower levels of political upheaval and we refer to these as *political unrest* and *instability*. These two variables were generated from a principal-components factor analysis of seven different variables indicating domestic political turmoil, obtained from the Banks (1999)

[1]For example, see Saleyhan and Gleditsch (2006) for a discussion of the regional clustering of civil war and the effects of refugee flows as one important factor in the diffusion of civil conflict. Note also the debate in the analysis of civil war, over the effects of ethnicity; see Fearon and Laitin (2003) or Elbadawi and Sambanis (2000) for results that minimize the effects of ethnicity on civil war.

dataset. Three of these variables—*strikes*, *riots*, and *demonstrations*—loaded on the first factor and the other four—*revolutions*, *coups*, *crises*, and *guerilla warfare*—loaded more heavily on the second factor. The *Political Unrest* variable is a combined indicator (factor score) for the first factor (strikes, riots, and demonstrations); this is essentially a measure of minor political disturbances. Somewhat more intense aspects of political tumult are indicated by the variable for *Political Instability*, which is the factor score for revolutions, coups, crises, and guerilla warfare.

The third dependent variable is a dichotomous indicator for the presence of *civil war* in a given country-year; the *Civil War* variable is coded one if there was a civil war and zero in the absence of a civil war. The civil war data were obtained from the PRIO Armed Conflict Dataset (Strand et al., 2004).[2] Another violent political consequence of state failure that we assess is *international war*. The indicator for *International War* is also a dichotomous variable, coded one for country-years with an interstate war and zero otherwise. These data were also obtained for the PRIO dataset (Strand et al., 2004).[3] The last consequence of state failure we consider is state collapse itself. To assess whether state failure leads to other state failures in nearby states, we include a binary variable for *State Failure* in a given country-year. States with a POLITY IV score of −77 in a given year are considered to have experienced a failure year and are assigned a value of one for that year; zero otherwise.

Covariates

The main covariates in these analyses include two measures for state failure that account for geography and a separate variable for the overall number of state failures in the system in a given year. The first independent variable is a *Distance-Weighted State Failure* measure that is obtained through the summation of all state failures in the system in a given year, where each one is weighted inversely by the log of the distance from the state in question:

$$\sum_{j=1}^{N_{t-1}} \frac{SF_{ijt}}{\ln(Distance_{ijt})} \qquad (1)$$

This continuous variable that measures distance-weighted state failure allows us to assess regional and second order effects of state collapse. The second independent variable measures the number of state failures that occurred in a given year in contiguous states. The value of the variable for *Collapse in a Contiguous State* ranges from zero to two. The purpose of including these two geographically adjusted measures of state failure is to discern the relative effects of contiguity and absolute distance on the diffusion of state collapse and its consequences. We expect that the closer a state is to a failed state, the more likely it is to experience the negative effects of state failure discussed above; we expect this relationship to be even stronger for contiguous states. The third covariate deals with the overall level of state failure in the system in a given year and is operationalized as the total number of states that failed in a particular year; the value of *State Failure in the System* ranges from zero to six. Given that state failure is believed to be regionally destabilizing, intuition might suggest that overall levels of state failure in the system would also have a destabilizing effect. We argue, however, that the diffusion of the effects of state fragility and collapse is a regional dynamic and does not operate at the systemic level. Thus, we do not expect the number of failures in the system to have a significant relationship with the

[2]The *Civil War* variable contains all conflicts designated "Type 3" in the PRIO Dataset on Armed Conflict. These are described as internal armed conflicts, and range in intensity from minor (with 25 battle deaths in a conflict) to intense (1000 battle deaths in a given country-year).

[3]All Type 2 conflicts (interstate armed conflict) in the PRIO Dataset on Armed Conflict.

TABLE 1 Summary statistics

Variables	Mean	Standard Deviation	Minimum	Maximum
Dependent variables				
Unrest ($N = 5560$)	0.030	1.039	-1.039	20.11
Instability ($N = 5560$)	0.009	1.013	−4.655	17.320
Civil war	0.136	0.343	0	1
State failure ($N = 5753$)	0.009	0.095	0	1
International war	0.057	0.231	0	1
Independent variables				
Distance-weighted state failure	0.239	0.208	0	1.297
Collapse in a contiguous state	0.050	0.226	0	2
State failures in the system	1.913	1.647	0	6
POLITY score	−0.598	7.662	−10	10
POLITY score squared	58.698	31.35	0	100
Population	0.029	0.101	0.0001	1.215
GDP	5.654	6.141	0.309	46.065
Openness	0.259	0.327	0	4.00

$N = 5755$ unless otherwise noted.
GDP = gross domestic product.

dependent variables. Moreover, the inclusion of this variable allows us to control for the effect of the total number of state failures in the system in our analysis of neighborhood and geographical effects.

The control variables in the analysis include levels of democracy, population, national income, and economic openness. Studies of civil war and political violence suggest that states that are "mixed regimes," neither highly democratic nor extremely autocratic, tend to be unstable and prone to violence. To test this proposition with regard to our five dependent variables, we include a variable for regime type, as measured by POLITY IV scores, as well as a squared term for POLITY scores (Marshall and Jaggers, 2004). The POLITY scale ranges from −10 (autocracy) to 10 (democracy). We expect a curvilinear relationship between democracy and the dependent variables, with the likelihood of state failure and its negative effects decreasing both at very high and very low levels of democracy. To capture basic socio-economic effects, we also control for population (measured in billions), national income (GDP per capita/1000), and economic openness (imports and exports as a proportion of national income).[4] The data on population figures and the two economic variables were obtained from the Expanded Trade and GDP Dataset (Gleditsch, 2002). Note that all independent variables are lagged one year to account for the effect of time.[5] Summary statistics are presented in Table 1.

[4]Collier and Hoeffler (2004) have shown that low national income is a major factor in the occurrence of civil war. Thus, while a variety of possible economic measures could be used, we have opted for more parsimony, and included only national income and economic openness.

[5]We chose to lag independent variables one year as this provides us with the most rigorous test of our hypotheses regarding diffusion and contagion of the consequences of state failure; the one-year lag reflects the "minimum incubation period" for the spread of the effects of state collapse. We expect longer lags to strengthen our findings.

TABLE 2 State failure and political unrest

Variable	Model 1	Model 2
Constant	0.077	0.07
	(0.05)	(0.05)
Distance-weighted state failure	0.23	—
	(0.48)	
Collapse in a contiguous state	—	0.18**
		(0.06)
State failures in the system	−0.04	−0.02
	(0.06)	(0.01)
POLITY score	0.01**	0.01**
	(0.003)	(0.003)
POLITY score squared	−0.002**	−0.001**
	(0.001)	(0.0006)
Population	2.07**	2.05**
	(0.27)	(0.27)
GDP	0.003	0.003
	(0.004)	(0.004)
Openness	−0.14*	−0.14*
	(0.06)	(0.06)
$\chi^2_{(7)}$	97.70	106.66

NT = 5560. Cell entries are random effects coefficient estimates; standard errors are in parentheses. One asterisk indicates $p < 0.05$ and two indicate $p < 0.01$, one-tailed.
GDP = gross domestic product.

Analyses and Results

We employ time-series cross-sectional data in our analyses and, therefore, we estimate a series of random-effects models to assess the likelihood of diffusion of the negative consequences of state failure to neighboring and nearby states. Given that these are panel data, the random-effects models allow us to take into consideration both the cross-sectional and temporal components of the data.[6] The models assessing the effect of the covariates on the continuous measures for political unrest and instability employ random-effects regression models, and analyses of the effects on the three dichotomous dependent variables (civil war, state failure, and international war) employ random-effects logit models (Hsiao, 2002). For each of our five dependent variables, we estimate two separate models: one that includes the distance-weighted state failure measure and the other contains the variable for the number of contiguous states that experienced failure in a given year. Below we outline the results of the models for each of our dependent variables.

With respect to the first dependent variable, our findings indicate that collapse in a contiguous state is positively related to unrest, although this effect does not hold for the

[6]We estimated the models using both random effects and fixed effects approaches, and the two approaches yielded substantively similar results. Under these circumstances, random effects models have the virtue of being more efficient, as well as preserving information from observations that fixed effects models must necessarily exclude (Hsiao, 2002). Alternative approaches to analyzing diffusion processes may include spatial lag and spatial error models. However, models for spatially referenced panel data (e.g., Franzese and Hays, 2007) are in their infancy, and we leave the use of those models for future work on the effects of state collapse.

TABLE 3 State failure and political instability

Variable	Model 1	Model 2
Constant	0.44	0.43
	(0.05)	(0.05)
Distance-weighted state failure	−0.17	—
	(0.50)	
Collapse in a contiguous state	—	0.12**
	—	(0.06)
State failures in the system	0.02	−0.009
	(0.06)	(0.008)
POLITY score	0.005*	0.005*
	(0.003)	(0.003)
POLITY score squared	−0.006**	−0.006**
	(0.001)	(0.001)
Population	−0.14	−0.15
	(0.29)	(0.28)
GDP	−0.009*	−0.009*
	(0.004)	(0.004)
Openness	−0.02	−0.02
	(0.06)	(0.06)
$\chi^2_{(7)}$	122.23	127.24

NT = 5560. Cell entries are random effects coefficient estimates; standard errors are in parentheses. One asterisk indicates $p < 0.05$ and two indicate $p < 0.01$, one-tailed.
GDP = gross domestic product.

distance-weighted measure (Table 2). Since this variable is a factor score with a mean of zero and a standard deviation of one, the coefficient of 0.18 indicates one-fifth of a standard deviation increase for each failed contiguous state. We find strong evidence for a curvilinear relationship between democracy and unrest. The size of the population is positively associated with unrest and economic openness has a negative effect. Unrest is not significantly affected by the other variables in the model. Instability levels are also positively related to state failure in a contiguous state, while the effect of the distance-weighted measure remains insignificant for this variable as well. Failure in a contiguous state increases instability by one-tenth of a standard deviation. Our expectations for a curvilinear relationship between democracy and instability are borne out in both models; hence we find evidence for the instability of mixed regimes (Table 3).

For both indicators of political turmoil (short of war), therefore, our models indicate that *contiguity*—not merely distance—plays an important role in the spread of the effects of state collapse. This reflects that the diffusion dynamics for these factors are best explained through the consideration of shared state borders. This is consistent with the findings of Iqbal (2007), which suggest that regional flows of refugees are best explained through consideration of contiguity.

Both models evaluating the effect of the covariates on incidence of civil war suggest that state failure increases the likelihood of civil war in nearby and neighboring states (Table 4). Incidence of state failure in a contiguous state increases the odds of a civil war by 45% and the odds of international conflict by 115%. The coefficients for both the distance-weighted state failure variable as well as the variable for Collapse in a Contiguous State are

TABLE 4 State failure and civil war

Variable	Model 1	Model 2
Constant	−2.47	−2.47
	(0.16)	(0.16)
Distance-weighted state failure	6.69**	—
	(2.23)	
Collapse in a contiguous state	—	0.37*
		(0.22)
State failures in the system	−0.67	0.15
	(0.28)	(0.03)
POLITY score	0.017	0.02
	(0.01)	(0.01)
POLITY score squared	−0.01**	−0.01**
	(0.002)	(0.002)
Population	4.22**	4.06**
	(0.98)	(0.98)
GDP	−0.007	−0.009
	(0.03)	(0.03)
Openness	0.09	0.11
	(0.26)	(0.26)
$\chi^2_{(7)}$	90.84	85.52

NT = 5755. Cell entries are random effects logit estimates; standard errors are in parentheses. One asterisk indicates $p < 0.05$ and two indicate $p < 0.01$, one-tailed.
GDP = gross domestic product.

positive and significant, although the former displays a higher level of significance. This suggests that civil war is a consequence of state failure that diffuses to states that are close by, regardless of contiguity. This implies the presence of salient diffusion processes for one of the most destabilizing factors associated with state failure: civil war. And the dispersion of this effect does not require a shared border. With respect to the control variables, the size of the population is also positively associated with civil war, although the curvilinear effect of democracy is less obvious than in the models assessing unrest and instability.

Another important finding of our analysis is a significantly positive effect of both the distance-weighted state failure measure and the indicator for collapse in contiguous states on the incidence of international war (Table 5). This is consistent with our expectation that state failure creates destabilizing conditions in a region, with possible incentives for violent behavior with other states (this is also consistent with the findings of Most and Starr, 1980; Starr and Most, 1983; 1985). The failed state itself may present an especially attractive target for invasion, thus providing opportunities and incentives for interstate armed conflict. This finding is particularly supportive of our expectation that state failure creates an unstable regional environment. The evidence that state failure is associated with both intra- and interstate conflict in neighboring and nearby states comports with the emphasis of international organizations and scholars on treating fragile and failing states as threats to international security. As in the previous models in our study, population is positively related to interstate conflict, and there is a minor negative effect of wealth. Note that there is no evidence of a curvilinear relationship between democracy and interstate conflict, which is consistent with the arguments in the democratic peace literature (see e.g., Russett, 1993;

TABLE 5 State failure and international war

Variable	Model 1	Model 2
Constant	−3.20	−3.15
	(0.24)	(0.23)
Distance-weighted state failure	7.91**	—
	(2.66)	
Collapse in a contiguous state	–	0.77**
		(0.27)
State failures in the system	−1.31	−0.36
	(0.34)	(0.05)
POLITY score	0.04	0.04
	(0.01)	(0.01)
POLITY score squared	0.003	0.003
	(0.003)	(0.003)
Population	3.77**	3.67**
	(0.90)	(0.89)
GDP	−0.04*	−0.04*
	(0.02)	(0.02)
Openness	−0.65	−0.55
	(0.45)	(0.44)
$\chi^2_{(7)}$	72.51	72.25

NT = 5755. Cell entries are random effects logit estimates; standard errors are in parentheses. One asterisk indicates $p < 0.05$ and two indicate $p < 0.01$, one-tailed.

GDP = gross domestic product.

Ray, 1998). Figure 1 presents the predicted probabilities of civil and international conflict as a function of distance-weighted state failure, while holding all other covariates at their means. A change in distance-weighted state failure from its minimum value of zero to its median value of 0.23 corresponds to an increase in the probability of civil war from 0.025 to 0.11. The smaller change in the probability of international war with increasing levels of state failure merely reflects the lower baseline probability of international conflict.

Although our analysis reveals that states geographically close to a failed state are likely to experience a civil war, and have a higher probability of getting involved in interstate conflict, that effect does not hold for state failure itself. Interestingly, neither of our models indicates that state failure is spatially contagious (Table 6). The only significant effect in the models assessing state failure as the dependent variable is the negative influence of GDP. The somewhat counterintuitive finding about the lack of contagion effects in state failure reflects that state collapse is distinct from other phenomena of political violence with which it is often associated, such as civil war. Our findings suggest that the regional threats posed by failing states are associated with instances of domestic and international violence of various magnitudes, but not a domino effect with respect to state collapse itself. State failure results from unique combinations of the characteristics of individual states and, therefore, certain domestic conditions may cause some states to fail while others remain alive (that is, make them "ready" for the diffusion of state failure). This is an internal, or domestic, argument. There is an external, or international, argument as well. Given a state

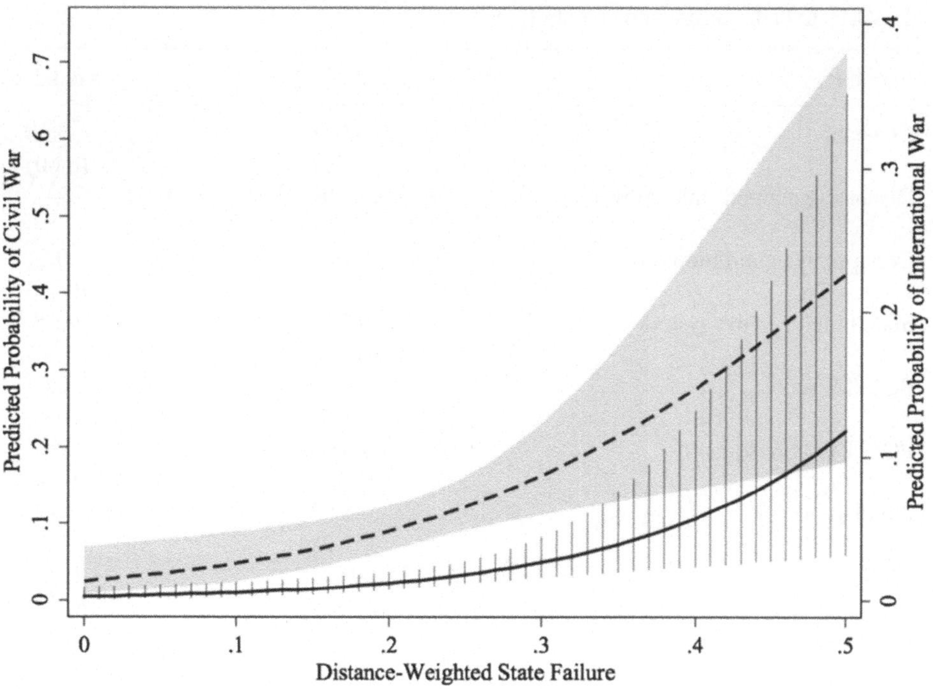

FIGURE 1 Predicted probability of war, by distance-weighted state failure.
Note: smooth line represents international war and dashed line represents civil war. Ranges are 95% point-wise confidence intervals.

failure in some country, a set of IGOs along with one or more major powers (and a number of regional powers) will often mobilize to provide political and material support to those nearby endangered states—especially given the perception and reality that various negative consequences of state failure will spread.

It should be noted that none of the dependent variables is affected by the overall number of state failures in the system. This finding is consistent with our argument that state failure negatively affects other states within close proximity, and as a result, can contribute to regional instability. The highest number of state failures in a given year is six, and it stands to reason that unless at least one of those failures is in a nearby or neighboring state, the likelihood of political violence does not increase. This emphasizes the importance of location, geography, and spatial dispersion in the understanding of issues related to domestic and international violence. If the diffusion of the consequences of state failure were a systemwide phenomenon, then these effects—such as civil wars—would rise as the number of failures increase. In that case, the variable for the total number of failures would capture that relationship and the distance-related measures would not add significantly to the explanatory power of the model. The results of our analyses, however, reveal the opposite pattern; the effects of variables for both distance-weighted state failure and failure in a contiguous state on the phenomena of interest are significant, whereas the global measure has no significant effect.

Taken together, the findings of our analyses imply that the consequences of state failure in terms of political violence do, in fact, geographically disperse and, therefore, may be destabilizing at the regional level. The consequences of state collapse that fall short of actual war (unrest and instability) spread only to contiguous states. Of particular note, however,

TABLE 6 State failure and state failure

Variable	Model 1	Model 2
Constant	−3.28	−3.30
	(0.60)	(0.60)
Distance-weighted state failure	−6.98	—
	(8.86)	
Collapse in a contiguous state	—	−0.21
		(0.76)
State failures in the system	0.92	0.05
	(1.12)	(0.10)
POLITY score	−0.05	−0.05
	(0.05)	(0.05)
POLITY score squared	−0.06**	−0.06**
	(0.008)	(0.008)
Population	−14.30	−14.17
	(17.28)	(17.33)
GDP	−0.43**	−0.42**
	(0.17)	(0.17)
Openness	0.83	0.83
	(0.36)	(0.36)
N	5753	5753
$\chi^2_{(7)}$	67.78	67.60

NT = 5753. Cell entries are random effects logit estimates; standard errors are in parentheses. One asterisk indicates $p < 0.05$ and two indicate $p < 0.01$, one-tailed.
GDP = gross domestic product.

is evidence that the diffusion of civil and interstate war is explained by both contiguity and distance. That is, these consequences are likely to spread to nearby states even in the absence of a shared border.

Conclusion

Our analyses indicate several major points regarding the regional effects of state failure. First, there are indeed spatial effects associated with collapsed states. Such states, which are generated by political instability and armed conflict, in turn generate subsequent unrest, instability, interstate, and civil war in their neighbors and near regions. Our analyses seem to indicate that the more “minor” possible consequences of state failure—political unrest and instability—spread to a lesser degree than more intense forms of violence—civil and interstate war. Following Most and Starr (1989), a quick review of some “stylized facts” indicates that this finding has face validity. Looking at all 73 of the country-years in which a state failure occurred (according to our measure of −77 on the POLITY scale), there is a new or ongoing civil war in almost 60% of those years. Similarly, civil war occurs in 55% of the sub-sample of −77 country-years in sub-Saharan Africa (while successful coups or coup attempts occur in only 18% of the sub-Saharan country-years) (McGowan, 2006).

Second, state failure itself is not contagious, although its major consequences are. This most extreme consequence of failure does not appear to diffuse either to contiguous neighbors, the regional neighborhood, or in the international system. What happens internally or externally that shields neighbors from this form of diffusion? As noted below, this is a key finding that raises a number of directions for future research.

Third, instability and political unrest appear to spread through contiguity, but not through a distance-weighted measure of state failure. On the other hand, the diffusion of civil war and international war is reflected in both measures of proximity. Hence, our findings support the position that distance-weighted measures and geographic contiguity tap related but not identical dimensions of proximity and its effects. And, as noted, diffusion thus appears to be regional as well as affecting first-order neighbors. This conclusion is further strengthened by the lack of a significant effect of the global measure for state failure on the phenomena of political violence under consideration. Although state failure in a nearby or neighboring state significantly increases armed conflict and other aspects of domestic political upheaval, the total number of states collapsing in the system in a given year has no effect.

Our study adds to a growing literature on the diffusion of violent conflict. One of the future projects to which this research points is a fuller examination of the diffusion of interstate conflict and of civil war, including the various ways the two forms of major conflict are related to one another. We have found civil war to be a major factor in the collapse of states. In turn, the current analyses indicate that civil war does diffuse to the neighbors/region of collapsed states.

Finally, there are a number of future directions for research specifically directed at state failure. As noted above, the exact mechanisms by which the conflict-oriented diffusion effects generated by failing states take place need to be investigated. Comparative case studies will also be needed for this purpose. This article has, indeed, paid most attention to the political/military consequences of collapse. There are, however, also a variety of nonpolitical consequences, including economic, social, and human security consequences of state collapse and violent conflict (see Iqbal, 2006). The investigation of economic and social consequences must take into account the finding here that democracy has a strong curvilinear effect on state collapse—with collapse least likely at both the highest and lowest levels of democracy.

Ultimately, we need to return to the policy concerns with which we opened this paper. The growing research agenda on various aspects of state failure must be aimed at devising policy prescriptions for managing and preventing the negative domestic and regional effects of state failure, as well as for preventing future failures. More broadly, such research will be instrumental in identifying effective ways for states and international organizations to address issues of instability, internal conflict, and development—especially during the highly vulnerable period of transition to democracy. Our research has raised a critical question: why doesn't state failure itself diffuse? This question relates directly to the various forms of early warning and intervention systems generated by international organizations and state governments. Scholars must look more closely at how IGOs (such as the UN and its various organizations, the World Bank, the IMF, or regional actors such as the EU) and states move to identify and deal with "at risk" states. We propose here that many such "at risk" or fragile states have been affected by the diffusion of the negative effects of state failure. Timely interventions might be an important factor in the lack of diffusion of state failure. What makes such interventions effective would be another focus of future research. And, intervention by whom? For economic and security support, some argue that the United Nations and other multilateral institutions, or the major powers, should take the lead (e.g., Mueller, 2004). Some findings on the democratic peace seem to indicate that stable and developed democracies might be crucial in helping "endangered" democracies avoid failure (e.g., Simon and Starr, 2000). In other areas, the lead may have been taken by

NGOs. Two things are clear: that the present analyses have raised a number of questions, and that much work remains to find the answers.

References

Banks, A. S. 1999. *Cross-national time-series data archive*. Binghamton, NY: Binghamton University—SUNY.

Beck, N., K. Gleditsch, and K. Beardsley. 2006. Space is more than geography: Using spatial econometrics in the study of political economy. *International Studies Quarterly* 50: 27–44.

Collier, P., and A. Hoeffler. 2004. Greed and grievance in civil war. *Oxford Economic Papers* 56(4): 563–595.

The Economist, March 5, 2005.

Elbadawi, I., and N. Sambanis. 2000. *How much war will we see? Estimating the incidence of civil war in 161 countries*. Geneva: World Bank.

Fearon, J. D., and D. Laitin. 2003. Ethnicity, insurgency, and civil war. *American Political Science Review* 97(1): 75–90.

Franzese, R., and J. Hays. 2007. Spatial-econometric models of cross-sectional interdependence in political-science panel and time-series-cross-section data. *Political Analysis* 15(2):140–164.

The Fund for Peace. 2006. *Failed state index*. Washington, DC: The Fund for Peace. http://www.fundforpeace.org/programs/fsi/fsindex.php.

Gleditsch, K., and M. D. Ward. 2002. Location, location, location: An MCMC approach to modeling the spatial context of war and peace. *Political Analysis* 10: 244–260.

Gleditsch, K. S. 2002. Expanded trade and GDP data. *Journal of Conflict Resolution* 46: 712–724.

Gleditsch, K. S. 2002. *All international politics is local: The diffusion of conflict, integration, and democratization*. Ann Arbor, MI: University of Michigan Press.

Gleditsch, K. S., and M. D. Ward. 2000. War and peace in space and time: The role of democratization. *International Studies Quarterly* 44(1): 1–29.

Goldstone, J. A., T. R. Gurr, B. Harff, M. A. Levy, M. G. Marshall, R. H. Bates, D. L. Epstein, C. H. Kahl, P. T. Surko, J. C. Ulfelder, Jr., and A. N. Unger, in consultation with M. Christenson, G. D. Dabelko, D. C. Esty, and T. M. Parris. 2000. *State failure task force report: Phase III findings*. McLean, VA: Science Applications International Corporation (SAIC).

Hsiao, C. 2002. *Analysis of panel data*, 2nd ed., New York: Cambridge University Press.

Iqbal, Z., and H. Starr. 2007. State failure: Conceptualization and determinants. Working paper, University of South Carolina.

Iqbal, Z. 2007. The geo-politics of forced migration in Africa, 1992–2001. *Conflict Management and Peace Science* 24: 105–119.

Iqbal, Z. 2006. Health and human security: The public health impact of violent conflict. *International Studies Quarterly* 50(3): 631–649.

Marshall, M. G., and K. Jaggers. 2004. *Polity IV project*. Integrated Network for Societal Conflict Research Program and Center for International Development and Conflict Management.

McGowan, P. J. 2006. Coups and conflict in West Africa, 1955–2004: Part II, empirical findings. *Armed Forces and Society* 32(2): 234–253.

Midlarsky, M. 1975. *On war*. New York: Free Press.

Most, B. A., and H. Starr. 1980. Diffusion, reinforcement, geopolitics, and the spread of war. *American Political Science Review* 74: 932–946.

Most, B. A., and H. Starr. 1989. *Inquiry, logic and international politics*. Columbia, SC: University of South Carolina Press.

Most, B. A., and H. Starr. 1990. Theoretical and logical issues in the study of international diffusion. *Journal of Theoretical Politics* 2(4): 391–412.

Mueller, J. 2004. *The remnants of war*. Ithaca: Cornell University Press.

O'Loughlin, J., and F. Witmer. 2005. Taking "geography" seriously: Disaggregating the study of civil wars. Paper presented at the conference on "Disaggregating the Study of Civil War and

Transnational Violence, University of California Institute of Global Conflict and Cooperation, La Jolla, CA., March 6–8.

Ray, J. L. 1998. Does democracy cause peace? *Annual Review of Political Science* 1: 27–46.

Russett, B. 1993. *Grasping the democratic peace: Principles for a post–Cold War world.* Princeton: Princeton University Press.

Salehyan, I., and K. S. Gleditsch. 2006. Refugees and the spread of civil war. *International Organization* 60(Spring): 335–366.

Sambanis, N. 2002. A review of recent advances and future directions in the quantitative literature on civil war. *Defence and Peace Economics* 13(3): 215–243.

Simmons, B., and Z. Elkins. 2004. The globalization of liberalization: Policy diffusion in the international political economy. *American Political Science Review* 98(1): 171–189.

Simon, M. V., and H. Starr. 2000. Two-level security management and the prospects for new democracies: A simulation analysis. *International Studies Quarterly* 44(3): 391–422.

Siverson, R. M., and H. Starr. 1991. *The diffusion of war.* Ann Arbor: University of Michigan Press.

Starr, H. 1994. Revolution and war: Rethinking the linkage between internal and external conflict. *Political Research Quarterly* 47(2): 481–507.

Starr, H. 2003. The power of place and the future of spatial analysis in the study of conflict. *Conflict Management and Peace Science* 20(1): 1–20.

Starr, H., and B. A. Most. 1983. Contagion and border effects on contemporary African conflict. *Comparative Political Studies* 16(1): 92–117.

Starr, H., and B. A. Most. 1985. The forms and processes of diffusion: Research update on contagion in African conflict. *Comparative Political Studies* 18: 206–277.

Strand, H., L. Wilhelmsen, and N. P. Gleditsch. 2004. *Armed conflict dataset codebook, Version 3.0.* Oslo: International Peace Research Institute.

UK Department for International Development. 2005. *Why we need to work more effectively in fragile states.* London: DFID.

World Bank. 2002. *World Bank Group work in low-income countries under stress.* Washington, DC: The World Bank Group.

What are the Preconditions for Turnarounds in Failing States?

LISA CHAUVET
PAUL COLLIER

Introduction

States can "fail" in various dimensions. The concept of state failure originates in political science, where the core function of the state has been seen as the need to maintain internal order. However, all modern states take on responsibilities that extend far beyond security. In particular, they undertake two socioeconomic functions: the provision of a wide range of public goods relating to health, transport, and education, and the regulation of private economic activity. In this paper, we focus on state failure, defined in this broader sense of the inadequate performance of socioeconomic functions. Economists routinely measure the outcomes of these functions, but in so doing they generally miss the *state*: what gets measured, as with the Millennium Development Goals, is the performance of the *country*. Yet an impoverished country with few opportunities may have poor socioeconomic outcomes, even if it has an effective state. One innovation of our analysis is to deploy a direct measure of state performance of these socioeconomic functions, rather than to infer performance from outcomes. If the state fails in the socioeconomic dimension, it is also more prone to fail in the security dimension: poverty and stagnation are significant risk factors in civil war (Miguel et al., 2004; Collier and Hoeffler, 2004a). In principle, however, the two dimensions are distinct, as is evident from the multitude of examples of states failing in one dimension but not the other.

The provision of public goods and the regulation of private economic activity are both achieved partly through institutions and partly through policies. Conceptually, institutions are the "rules of the game" (North, 1990), and are slow-changing. Acemoglu et al. (2001) claim to have shown that institutions are persistent over centuries and that they are fundamental to economic performance. They suggest that short-term measures of policy change, which abstract from institutions, cannot provide an adequate measure of state performance. However, their interpretation of their results as demonstrating institutional persistence has been challenged. In the typical failing state, institutions may be highly fluid because the state lacks the capacity to make credible commitments to any institutional stability. Hence, the distinction between institutions and policies may be blurred. For example, on paper, the justice system may look very like that prevailing prior to independence, but its practical functioning may be very different and subject to rapid change. Any reasonable measure of state performance within the range of state failure is therefore likely to be some composite assessment of the quality of institutions and the quality of policies.

The concept of state failure is used chiefly by international agencies and the governments of those states contemplating some form of assistance. Thus, the assessment of state performance should ideally be one that they themselves have made. We make use of the most well-established such assessment, the Country Policy and Institutional Assessment (CPIA) of the World Bank. As its name implies, the CPIA is a composite of policies and institutions, assessed on twenty different aspects of performance. One important advantage of the CPIA is that is it indeed intended to be a measure of *state* performance. It is, in principle, entirely possible for an impoverished country with very poor socioeconomic outcomes to get the maximum rating on this measure, as long as the state is performing its public goods and regulatory functions as well as is possible under difficult conditions.

State performance can, in principle, be ranged along a continuum, and the CPIA is such a measure, However, the concept of state "failure" implies some threshold below which performance is in some sense radically unsatisfactory. The conversion of continuous variables to discrete variables through the adoption of a threshold is common to many important, policy issues. It has the advantages of concentrating on that part of the range judged to be most important and of generating clear events when the threshold is crossed. The price paid is that the precise location of a threshold is inevitably arbitrary, so that it is important to investigate the robustness of any results to reasonable variations in the positioning of the threshold. We convert the CPIA into a discrete variable by imposing a threshold.

The concept of state failure is rooted in the potential for external action. When the nature of the failure is an inability to maintain internal security, the idea of international trusteeships enters active discussion (Fearon and Laitin, 2004; Krasner, 2004). However, for the type of state failure considered here, the more pertinent forms of international intervention are likely to be the provision of financial assistance and qualified people, i.e., aid and technical assistance. Such assistance can potentially have two very different rationales: palliative and ameliorative. Assistance has a palliative effect if it succeeds in delivering public services that would not otherwise be delivered by the state itself. It has an ameliorative effect if it succeeds in improving the policies and institutions of the state so that, in due course, the state itself is better able to perform these functions. The palliative and ameliorative functions of assistance may complement or offset each other. Conditional aid was an attempt to make the two roles complementary: directly money would enhance service delivery, which, indirectly through conditionality, would force improvements in policies and institutions. Similarly, technical assistance was intended both to supplement the capacity of the civil service by directly performing some functions and, through training, to build that capacity. In this paper, we abstract from the palliative effects of aid and technical assistance and focus entirely upon their ameliorative effects.

While it is clear that in principle each could have ameliorative effects, much of the literature on aid effectiveness raises doubts as to its practical efficacy. One strand of the aid-effectiveness literature suggests that the absorptive capacity for aid might be limited in the conditions of weak policies and institutions that characterize failing states (Burnside and Dollar, 2000; Collier and Dollar, 2002). However, even if failing states can absorb aid, the effect may be purely palliative. A substantial case study and econometric literature on donor conditionality has concluded that it was largely ineffective in inducing policy reform. A good example of the case study literature is Devarajan et al. (2001), who examined ten reform episodes to a common framework. The key econometric contribution is probably that of Dollar and Svensson (2000), who found no significant causal relationship between aid programs and the reform of governance and policy. The incentive effect of aid on reform is indeed highly doubtful. In economic terms, there are two opposing effects. Linking aid to reform indeed creates a favorable "substitution effect," making reform more attractive for a government (Sachs, 1994). However, this is offset by an "income effect," i.e., the increase in aid makes change less necessary (Rodrik, 1996). Perhaps more important, in psychological terms, the infringement of freedom associated with conditions induces "reactance," whereby governments attempt to reestablish their freedom by doing the opposite of what the conditions require (Collier, 2001).

The incentive effect of aid on reform, however, is by no means the only way in which aid can influence policy and institutions. Aid can build capacity, either directly or indirectly. It can expose governments to new ideas. It can free governments from crisis management, enabling them to think about longer-term strategies. Such effects might be particularly important at the very bottom of the economic spectrum. Chauvet and Guillaumont (2004) estimate economic policy regressions and indeed find that when policies are initially very poor, aid has a positive impact on them. Given that the general result of Dollar and Svensson (2000) is surely robust, it is worth investigating whether aid promotes policy and institutional reform within the range of failing states.

Thus, the purpose of this paper is to investigate empirically the scope for aid and technical assistance to enable failing states decisively to turn around, achieving and sustaining radically improved policies and institutions. Specifically, we investigate whether aid provided prior to reform promotes such turnarounds.

In Section 2, we define more precisely the concept of a sustained turnaround. In Section 3, we turn to the theoretical and empirical preconditions for such a turnaround. Section 4 provides a discussion of the implications for policy. Finally, Section 5 summarizes our findings.

Defining Sustained Turnaround

In general terms, the phenomenon we wish to analyze is clear enough. However, we need both a measure of economic policies and institutions, and precise definitions of what is meant by a "failing state" and by a "sustained turnaround." As a measure of policy, institutions, and governance we will use the Country Policy and Institutional Assessment (CPIA) of the World Bank. This is an internal World Bank standardized subjective rating system done annually for all countries that are borrowers from the World Bank.[1] Twenty different aspects of policy, institutions, and governance are rated using criteria that grade each of them on an ascending scale of 1 to 6 (see Appendix 1 for the definition of the CPIA). These are averaged by the Bank to form the rating that we use. The CPIA has the considerable advantage that it is the only such rating continuously available since 1977, explicitly designed to be comparable across countries. The major disadvantage is that it is a subjective index that incorporates judgments of World Bank economists, creating the potential for two types of error. First, some of the judgments reflect economic analysis that is professionally contentious. Fortunately for our purposes, the contentious judgments arise overwhelmingly in the higher reaches of the CPIA. To illustrate outside the range of rated countries, it is evident that significant differences exist between France, the U.S., the U.K., and Japan, but there would be no professional consensus as to which of these countries had the "better" policies, institutions, and governance. Our focus on failing states takes us to the bottom end of the CPIA range. Here there is little controversy. There is an overwhelming professional consensus that countries such as Angola, Burundi, and the Central African Republic have had bad policies, institutions, and governance. A second potential for error is that, despite the overview procedures, some changes in the CPIA may reflect nothing more than changes in staffing. We guard against this by devising criteria that are not dependent upon small changes in the score. Thus, for our purposes, the CPIA is a reasonable index for the phenomenon we wish to classify. As with all continua, there will, however, be issues at the margin of what is, and what is not, a failing state. To specify a turnaround we need to answer the questions "from what?" and "to what?" It is to these issues that we now turn.

For our purposes, to be a failing state the country must have a low income. We use as an income cutoff that the country should have been classified as a LIC for at least one year by the World Bank (World Development Reports, 1977–2004). Evidently, not all low-income countries are failing states. We adopt as a cutoff for weakness in policy, institutions, and governance, a level of the CPIA no better than 2.5, which is a very low rating.

However, many countries occasionally meet these criteria and yet are not appropriate for the category of turnaround. This is because their period of difficulty reflects only a temporary crisis into which they plunge and from which they rapidly rebound. A failing state is not simply a country in a brief spell of crisis. In particular, a country facing a crash in its policies, institutions, and governance may be able to recover rapidly because expectations are not set, and because those political groups that benefit from the new situation have not yet mustered the power to prevent change. The concept of a failing state includes some notion that the problem of weak policies, institutions, and governance is not just of the moment, but has some persistence. We therefore add the requirement that the CPIA should have been under the threshold of 2.5 for at least 4 consecutive years. Countries that meet the other criteria but not this one are termed "recoveries" rather than failing states.

Thus defined, we have the pool of all country episodes from which a turnaround from being a failing state might occur. From within this pool, we now define what we mean by a turnaround.

Evidently, to be a turnaround, policies, institutions, and governance must improve above the threshold for a failing state. However, this alone is clearly insufficient. A country

[1]Because it has been used to allocate the Bank's concessional lending (IDA), the details of the ratings have been kept confidential from the countries concerned, although it has been published in quintiles for some years. We understand that the Bank's Board has decided in 2006 to release the detailed information.

whose CPIA improved from 4 consecutive years of being 2.49 to reach 2.51 would meet the letter of such a condition but would not have had a significant improvement. We introduce the additional requirement that the CPIA should attain at least 3.5. As a robustness check, we also test an alternative threshold requirement: the CPIA starts below 2.5 and reaches at least 3.

Momentary improvements, however large, are evidently not sustained turnarounds. We therefore need criteria for sustainability. It may seem obvious that the criterion for sustainability is simply that the reform should have persisted to the present. However, this would be a poor criterion. To quote a famous African saying, "no condition is permanent." Taking a specific example, Indonesia had a dramatic turnaround from the late 1960s, at which time it was a classic failing state, to become a star performer of the 1980s. Yet, during the late 1990s it suffered a collapse along with much of East Asia. It seems to us unreasonable to see this crisis of the late 1990s as being the result of failures in the design of the turnaround. It is surely more reasonable to think of the turnaround as having been successful in producing sustained improvement, with the crisis of the late 1990s being attributable to some intervening event or process subsequent to the turnaround. By "unsustained," we wish to capture only those turnarounds where the subsequent relapse was sufficiently close to the turnaround that the reasons for it could sensibly be attributed, in large part, to weaknesses in the turnaround itself. We therefore deem a turnaround to have been sustained if the CPIA remains above 3 for at least 2 years after the turnaround is achieved. A further reason for setting such a limit to the requirement of sustainability is that otherwise relatively recent turnarounds look much more successful than earlier turnarounds, which have simply had more time to be reversed.

Table 1 lists the failing states that never implemented a sustained reform as well as those that managed to exit this category following a sustained turnaround.

Preconditions for Sustained Turnaround

We now have precise criteria both for a country in a position to have a turnaround, and for what constitutes a sustained turnaround. The issue we investigate in this section is what preconditions significantly increase the prospects of such a turnaround.

An empirical investigation of turnarounds must be grounded in a theory of reform. By definition, a failing state is not performing to a standard that provides its citizens with an adequate environment in terms of either social provision or the economic activity. In some sense, the society as a whole is losing out. Yet also, by our definition, this failure is not merely momentary: it is a persistent equilibrium. Logically, there are two broad sets of explanations for the persistence of a socially dysfunctional situation. One possibility is that it is individually optimal because the cooperation required to reach socially superior outcomes is infeasible. Alternatively, the society does not have the capacity to recognize the dysfunctionality, despite its being evident to informed outsiders.

The most thoroughly explored instance of the former problem is associated with natural resource abundance: the "resource curse" (Sachs and Warner, 1995; 2001). Essentially, the existence of valuable natural resources increases the payoff for subsets of the society to hold power and use it to their own advantage. The most stark illustration can be found in the work of Ross (2001), who found that oil wealth reduces the likelihood of democracy. However, the problem is more general: even democracies in resource-rich societies may be unable to generate cooperative outcomes. For example, Lane and Tornell generate a theoretical prediction of a "voracity effect," whereby resource rents so reduce the capacity for cooperation that they immizerize (Lane and Tornell, 1996; Tornell and Lane, 1999).

TABLE 1 List of failing states

Failing state 1977–2004 No sustained reforms	Failing states that managed to implement a sustained reform	Date
Angola	Albania	1991–1995
Cambodia	Bangladesh*	1977–2000
Chad	Benin*	1987–2002
Comoros	Burkina Faso	1987–1994
Congo, Dem. Rep.	Burundi	1983–1988
Congo, Rep.	Central African Rep.*	1977–1989
Equatorial Guinea	Côte d'Ivoire	1992–2000
Guinea	Egypt*	1988–2000
Guinea-Bissau	Ethiopia*	1991–2000
Haiti	Ghana*	1982–1987
Laos	Guyana	1987–1996
Liberia	Honduras*	1989–1995
Mauritania	Indonesia*	1977–1987
Mozambique	Lesotho	1987–1998
Myanmar	Madagascar	1978–1991
Niger	Nepal*	1985–2003
Nigeria	Nicaragua*	1989–1996
Iles Salomon	Pakistan*	1979–1990
Sao Tome & Principe	Rwanda	1994–2003
Sierra Leone	Tajikistan	1997–2003
Somalia	Togo*	1980–1987
Sri Lanka	Uganda	1986–1998
Sudan	Ukraine	1999–2003
Tanzania	Vietnam	1988–1991
Turkmenistan	Zambia	1990–2000
Uzbekistan		

Sustained reforms are defined as an increase in the CPIA from 2.5 to 3.5, sustained for at least 2 years.

*Sustained turnarounds included in our baseline regression.

Ades and Di Tella (1999) even show that natural resource rents may increase corruption. Melhum et al. (2006) show that the impact of natural resource rent on growth depends on the quality of institutions. Notably, natural resources would have a negative impact on income growth when institutions are grabber-friendly. The resource curse literature thus predicts that resource rents would increase the persistence of being a failing state. Since resource rents are observable, this is a testable proposition. Although resource rents are probably the most salient impediment to collective action, they are by no means the only one, and we will consider others in our empirical analysis.

The alternative family of explanations for the persistence of a dysfunctional equilibrium is that there is insufficient "social knowledge" to build a consensus that reform is mutually beneficial. Many desirable economic policies, such as trade liberalization or fiscal caution, are highly counterintuitive. Similarly, the key institutional structures necessary to overcome the "time-consistency" problem require that the state limit its own power, a concept that is sufficiently sophisticated that it has only recently been properly understood in developed societies. More fundamentally, the basic mechanisms of political accountability depend upon citizens being able to distinguish between those outcomes that are the responsibility of government and those that are the result of chance. In his important study of political accountability, Besley (2006) shows how it is affected by the educational level of citizens: a low average level of education induces unprincipled behavior in politicians. Following Besley, we adopt a measure of the extent of secondary education as one of our proxies for differences in "social knowledge." Our other measure is the size of the population. This is meant to capture scale economies in knowledge. The standard example of such scale effects is the number of innovators in the society: larger societies tend to have more innovators. However, scale also offers more mundane economies, such as a market sufficiently large to support specialized media, most notably a financial press. Societies that have few well-educated people, whether because only a small proportion of the society is educated, or because the population is small, or due to a combination of these characteristics, may be at a disadvantage in diagnosing that policies and institutions are dysfunctional, and in designing a strategy of reform.

Within these two families of theories of reform, how might external assistance have effects? Technical assistance clearly has some potential to augment social knowledge. Its core purpose is to transfer technical information, most commonly based on some notion of "best practice" from other societies. However, potentially technical assistance can confuse, either by being inappropriate for local circumstances, or more importantly by inducing "reactance," the sense by local actors that externally imposed ideas should be resisted. A classic example of reactance is the response of the Nigerian intellectual community to the Structural Adjustment Program advised by the World Bank following the 1986 collapse in the world oil price. For many years following that program, there was overwhelming domestic opposition to economic reform, despite highly dysfunctional economic governance. Financial aid can also have ambiguous effects. The ambiguities associated with conditionality have already been discussed. A further potential difficulty arises from the political economy of crises. Crisis provides the impetus to overcoming the impediments to collective action, a finding common to major collections of case studies of the reform process (Little et al., 1993; Deverajan et al., 2001). By cushioning crises, aid can thus potentially inadvertently delay reform.

We now turn to our empirical method. We use probit regressions to estimate the probability of a turnaround, year by year, among all potential turnaround countries. We proceed as follows: First we identify a robust baseline specification. Our approach is empirical: based on specification tests we identify what determines the start of sustained turnarounds in failing states. As discussed, we distinguish between two components of aid, technical assistance and other aid, both being lagged. Our source for these data is the reporting system of the Development Assistance Committee (DAC) of the OECD. This reporting system leaves much to be desired. However, while it most certainly contains errors and inconsistencies, it may nevertheless have some informational content.

Then in a second step, we need to surmount the problem of the endogeneity of aid. Aid is allocated purposively. Donors seek out information so as to direct aid to those situations where it stands the best chance of being helpful. If donors try to target situations that are ripe for improvement, it is intrinsically difficult to determine the direction of causality. We overcome this problem by instrumenting for aid. However, the issue of endogeneity is more likely to be an issue when turnarounds have already started (Chauvet and Collier, 2006). Indeed, it is very difficult for donors to anticipate turnarounds, and even if they have sometimes done so, they are unlikely to have differentiated between technical assistance and other aid.

The Baseline Specification

Our probit regression investigates the chances of a sustained turnaround from being a failing state. Since the issue of aid endogeneity will be tackled in a second step, we introduce aid and technical assistance lagged one-period. We start with annual observations: each year a failing state has some probability of starting an improvement in policies, institutions, and governance that culminates in a sustained transformation. Once it has started on a sustained turnaround it drops from our sample since it cannot start a further turnaround.

In Table 2, we present our core regression. All data and variables are presented in detail in Appendix 2.[2] Consistent with our overview of theories of reform, our proxies for the resource curse and for social knowledge are all significant. Specifically, we measure the value of resource rents relative to GDP, the proportion of the population with secondary education, and the size of the population. Countries benefiting from large resource rents tend to be less likely to achieve sustained turnarounds. Countries with a higher proportion of their people who have secondary education, and countries with larger populations, are significantly more likely to achieve sustained reforms. We now turn to the two external assistance variables, which at this preliminary stage are not instrumented. Aid in the form of financial assistance has a negative coefficient, although this is not significant. This is at least qualitatively consistent with the adverse effect of finance through natural resource rents. In contrast, technical assistance, lagged one period, has a significantly positive effect.

We perform specification tests using several political and economic variables: income per capita, democracy, civil war, post-conflict situations. None of them is significant. In Table 3, we test the robustness of our baseline specification using an alternative dependent variable (a threshold from 2.5 to 3 instead of 2.5 to 3.5). We also test for robustness by controlling for country random effects. The results are stable. Finally, in Table 4, we test the robustness of our baseline specification to the use of 4-year averages instead of annual data. Once again, the results are reasonably stable.

The Endogeneity of Aid and Technical Assistance

We adopt a set of instruments pioneered by Tavares (2003) that are now common in the literature. Essentially, the idea is that a substantial component of a country's aid receipts is determined not by its own current circumstances but by the characteristics of donors. For example, Ethiopia is likely to get a relatively large amount of aid from Italy, and Cote d'Ivoire is likely to get relatively a lot of aid from France. If the Italian aid budget goes up and the French aid budget goes down, Ethiopia is likely to get an increase in its aid receipts relative to Cote d'Ivoire. So we use as instruments the total aid budget of the main bilateral donors, as well as some measures of cultural distance between these donors and the recipients: dummy variables equal to one when recipient and donor countries share a common language, dummy variables equal to one if 30% or more of the population belong to a religious group in the donor and in the recipient country. We investigate a set of around 20 instruments (and 15 more when we include interactions between distance variables and total aid budgets). We keep only a few of them, those which are significant: the total budget of aid of Japan (does not vary across countries), a dummy for the Catholic religion, and a dummy for being English-speaking (these evidently not varying across time). We supplement these instruments with aid and technical assistance lagged twice. Table 5 presents the results using both annual data and 4-year averages. The last three rows of Table 5 provide tests for the validity of the instruments (obtained from TSLS estimations): Hansen over-identification test and partial R^2 of excluded instruments.

[2]Countries included in the sample are listed in Table 1.

TABLE 2 Determinants of the start of sustained turnarounds, probit estimations, annual data, 1973–2001

Start of a sustained turnaround (from 2.5 to 3.5)	(1)	(2)	(3)	(4)	(5)	(6)
Secondary schooling	0.161 (2.50)**	0.152 (2.53)**	0.162 (2.47)**	0.162 (2.49)**	0.163 (2.58)***	0.194 (2.97)***
Resource rent (% GDP)	−4.565 (1.76)*	−5.347 (1.74)*	−4.750 (1.71)*	−4.470 (1.71)*	−4.568 (1.77)*	−5.264 (1.85)*
Ln population	0.298 (1.89)*	0.360 (1.86)*	0.290 (1.86)*	0.281 (1.77)*	0.296 (1.78)*	0.285 (1.81)*
Aid (% GDP), lagged	−0.058 (1.62)	−0.060 (1.61)	−0.066 (1.62)	−0.060* (1.69)	−0.058 (1.59)	−0.047 (1.55)
Tech. assist. (% GDP), lagged	0.219 (1.72)*	0.238 (1.74)*	0.233 (1.73)*	0.235 (1.78)*	0.218 (1.70)*	0.234 (1.88)*
Ln real income per capita, lagged		0.249 (0.93)				
Postconflict, years 1 to 4			0.499 (0.90)			
At war (dummy = 1)				0.222 (0.55)		
Duration leader in office					0.002 (0.12)	
Democracy (0–10)						−0.102 (1.62)
Constant	−4.884 (3.19)**	−7.218 (2.13)*	−4.808 (3.14)**	−4.787 (3.19)**	−4.885 (3.18)**	−4.805 (3.19)**
Observations	344	341	338	338	342	342
Number of sustained turnarounds	12	12	12	12	12	12

Student t-test corrected for heteroscedasticity in parentheses. ***significant at 1%; **significant at 5%; *significant at 10%.

TABLE 3 Robustness tests: alternative dependent variable and estimation method (annual data, 1973–2001)

	Probit		Random effects probit	
Start of a sustained turnaround	From 2.5 to 3.5 (1)	From 2 to 3 (2)	From 2.5 to 3.5 (3)	From 2 to 3 (4)
Secondary schooling	0.161 (2.50)**	0.121 (1.70)*	0.218 (1.81)*	0.168 (1.56)
Resource rent (% GDP)	−4.565 (1.76)*	−5.211 (1.90)*	−6.004 (1.44)	−7.080 (1.69)*
Ln population	0.298 (1.89)*	0.390 (2.38)**	0.384 (1.86)*	0.520 (2.37)**
Aid (% GDP), lagged	−0.058 (1.62)	−0.038 (1.05)	−0.080 (1.35)	−0.047 (0.80)
Technical assistance (% GDP), lagged	0.219 (1.72)*	0.308 (2.16)**	0.346 (1.60) ($p = 0.11$)	0.472 (2.26)**
Constant	−4.884 (3.19)***	−5.797 (3.56)***	−5.960 (2.79)***	−7.335 (3.19)***
Observations	344	323	344	323
Number of countries	22	22	22	22
Number of sustained turnarounds	12	14	12	14
Significance of random effects (1)			0.115	0.075
Specification tests (p-value):				
Income	0.351	0.565	0.705	0.962
Postconflict	0.369	0.229	0.445	0.277
Civil war	0.583	0.259	0.637	0.352
Democracy	0.104	0.506	0.175	0.506
Duration in power	0.907	0.420	0.504	0.878

Student t-test corrected for heteroscedasticity in parentheses. ***significant at 1%; **significant at 5%; *significant at 10%. (1) Probability that the panel estimator is not different from the pooled estimator.

TABLE 4 Robustness tests: alternative dependent variable and estimation on 4-year sub-periods, 1974–2001

	Probit		Random effects probit	
	From 2.5 to 3.5	From 2 to 3	From 2.5 to 3.5	From 2 to 3
Start of a sustained turnaround	(1)	(2)	(4)	(5)
Secondary schooling	0.243 (2.78)***	0.164 (1.83)*	0.356 (1.94)*	0.268 (1.64)*
Resource rent (% GDP)	−1.109 (1.28)	−0.403 (0.45)	−1.768 (0.98)	−0.951 (0.59)
Ln population	0.432 (2.38)**	0.484 (2.72)***	0.550 (2.04)**	0.654 (2.31)**
Aid (% GDP), lagged	−0.129 (2.04)**	−0.070 (1.48)	−0.165 (1.55)	−0.086 (1.00)
Technical assistance (% GDP), lagged	0.390 (1.66)*	0.424 (2.03)**	0.587 (1.69)*	0.673 (1.99)**
Constant	−5.755 (3.09)***	−6.252 (3.39)***	−7.319 (2.52)**	−8.401 (2.72)***
Observations	108	103	108	103
Number of countries	24	24	24	24
Number of sustained turnarounds	14	16	14	16
Significance of random effects (1)			0.114	0.075
Specification tests (p-value)				
Income	0.448	0.935	0.610	0.954
Postconflict	0.156	0.254	0.299	0.441
Civil war	0.818	0.528	0.952	0.696
Democracy	0.073	0.268	0.111	0.292
Duration in power	0.945	0.360	0.492	0.790

Student t-test corrected for heteroscedasticity in parentheses. ***significant at 1%; >**significant at 5%; *significant at 10%. (1) Probability that the panel estimator is not different from the pooled estimator.

TABLE 5 Instrumentation of aid and technical assistance: estimation on 4-year sub-periods and annual data, 1973–2001

	Annual data			4-year averages		
IV Probit Maximum Likelihood Estimation	Start of a ST (1)	ODA (2)	TA (3)	Start of a ST (4)	ODA (5)	TA (6)
Constant	−5.162 (3.63)***	1.181 (0.54)	2.655 (4.65)***	−4.757 (3.37)***	1.491 (0.29)	3.171 (3.41)***
Secondary schooling	0.153 (2.56)**	−0.427 (2.40)**	−0.079 (2.61)***	0.156 (1.91)*	−0.967 (1.97)**	−0.189 (2.61)***
Resource rent (% GDP)	−5.074 (1.84)*	−6.846 (2.51)**	−1.349 (1.90) *	−4.250 (1.72)*	7.886 (1.39)	−2.779 (2.05)**
Ln population	0.342 (2.30)**	0.0004 (0.00)	−0.218 (4.33)***	0.403 (2.94)***	0.306 (0.56)	−0.241 (2.75)***
Aid (% GDP)	−0.094 (2.67)***			−0.186 (4.37)***		
Technical assistance (% GDP)	0.321 (1.98)**			0.589 (4.12)***		
Excluded instruments						
Aid (% GDP), lagged twice		0.707 (6.87)***	0.059 (2.65)***		0.359 (2.67)***	0.036 (0.95)
Technical assistance (% GDP), lagged twice		−0.127 (0.38)	0.466 (5.07)***		0.088 (0.19)	0.268 (2.28)**
Same religion as France		0.716 (1.06)	−0.369 (2.77)***		1.891 (1.30)	−0.250 (0.94)
Same language as UK		−0.882 (1.42)	−0.573 (3.91)***		−1.233 (1.08)	−0.924 (4.27)***
Total budget of aid of Japan		0.0005 (3.01)***	0.0001 (3.21)***		0.001 (2.25)***	0.0002 (3.34)***
Observations	343			104		
Number of countries (sustained turnarounds)	22 (12)			24 (14)		
Wald test (*p*-value)	0.012			0.000		
Log pseudolikelihood	−1515.5			−483.5		
Test of exogeneity (*p*-value)(1)	0.147			0.004		
Hansen J statistics (2)	0.956			0.576		
Partial R^2 of excluded instruments (ODA) (2)	0.59			0.34		
Partial R^2 of excluded instruments (TA) (2)	0.62			0.47		

Estimations using ivprobit routine of Stata. Start of ST defined as improvement in the CPIA from 2.5 to 3.5. Student *t*-test corrected for heteroscedasticity in parentheses. ***significant at 1%; **significant at 5%; *significant at 10%. (1) Probability that aid and technical assistance are exogenous. (2) Obtained from TSLS regressions.

Consistent with the probit results, we find that technical assistance prior to turnaround has a significantly positive effect on the prospects of turnaround: providing advice and implementation capacity seems to matter in failing states. The effect of aid other than technical assistance is quite different. We find that aid as finance has significantly adverse effects on the chances for a sustained turnaround. Whereas aid was not significant in probit estimations, it becomes significant once instrumented.

Policy Simulations

From the first regression of Table 3, we can compute the average probability of a failing state having a sustained turnaround. Overall, we find that the probability of a sustained turnaround starting in any year is very low, at 1.85%. Countries are therefore likely to stay as failing states for a very long time. Indeed, given the probability, the mathematical expectation of the duration is 54 years. Note that this may slightly exaggerate the duration because it considers only exits due to improvement in policies, institutions, and governance. A country can also cease to be a failing state because of a rise in income that takes it above the low-income threshold. While on average, very poor policies, institutions, and governance make such a rise in income unlikely, sometimes countries have good fortune. For example, Equatorial Guinea is not, on our criteria, currently a failing state because its income level is too high due to an oil discovery. While our estimate of persistence abstracts from such occurrences, growth in the presence of failing state-type policies, institutions and governance is unlikely to be significantly poverty reducing, as indeed is indicated by Equatorial Guinea.

Extra Technical Assistance Prior to Reform

We first consider a small increase in technical assistance delivered prior to an incipient turnaround. On average, failing states have received technical assistance equal to 2.1% of their GPD. We therefore simulate the effect of increasing this by one percentage point to 3.1%. The effect it to raise the annual probability of achieving a sustained turnaround from 1.85% per year to 3.1% per year. So, even if the average probability is still very low, it is almost doubled when technical assistance is increased by 1%. This substantially shortens the expected delay in decisively exiting the category of failing state, from 54 years to 32 years.

An Expansion in Secondary Education Prior to Reform

We next simulate the effect of increasing the proportion of the population with secondary education from the mean found in failing states, 2.13%, to 3.13%. This would raise the probability of reform from 1.8% per year to 2.7% per year. This also substantially reduces the expected duration of state failure, from 54 years to 37 years. If the policy of expanded secondary education is combined with that of enhanced technical assistance, the period of state failure is shortened dramatically, from 54 years to only 23 years, the annual probability of exit now rising to 4.4%.

Increases in External Finance

Finally, we consider two forms of additional external finance, rents on resource exports and aid other than technical assistance. An expansion in the natural resource rent is detrimental to reform. When resource rent increase from 5.6% of GDP to 6.6% of GDP, the average probability goes from 1.8% to 1.65%. During the present commodity boom, resource rents have increased very substantially, and so we also investigate the consequence of a doubling of these rents. In response to a doubling of rents, the duration of state failure also virtually doubles to 104 years. An expansion in aid as finance has a very similar impact on the probability of reform as an increase in resource rent. When aid increases from 6.94% of GDP (the average of the sample) to 7.94% of GDP, the average probability goes from 1.8% to 1.59%, extending the period of state failure by nearly a decade to 63 years. While each of these effects is only marginally significant, the fact that these two different sources of government finance have the same chilling effect on the impetus to reform increases our confidence in the results.

Conclusion

In this paper we have proposed an economic definition of a failing state: a low-income country with very weak economic policies, institutions, and governance. We have found such conditions to be highly persistent: with mean characteristics, the mathematical expectation of the time until a decisive change is well established is around 54 years.

Although failing states arouse concern, there is a degree of pessimism as to what can be done about them. The high costs and controversy surrounding U.S. military intervention in Iraq and Afghanistan raise doubts as to the future viability of military solutions. A recent political literature is flirting with the revival of international trusteeship, or colonialism mark 2. Aid tends to be dismissed as being not up to the task, or even as part of the problem. In part, this reflects current thinking that aid is most effective where policies and institutions are already reasonable, while past attempts to improve policies through donor conditionality are judged to have failed. Consistent with these priors, we have found some evidence that windfall external finance prior to reform, whether from additional rents on resource exports, or aid, significantly retards reform. However, more targeted forms of aid can be effective in promoting policy turnarounds. Technical assistance significantly and substantially promotes reform, and, subject to the problem of fungibility, aid for the expansion of secondary education also promotes reform.

References

Ades, A., and R. Di Tella. 1999. Rents, competition, and corruption. *American Economic Review* 89(4): 982–993.

Acemoglu, D., S. Johnson, and J. A. Robinson. 2001. The colonial origins of comparative development: An empirical investigation. *American Economic Review* 91(5): 1369–1401.

Barrett, D. B. (ed.). 1982. *World Christian encyclopedia.* Oxford University Press, Oxford.

Barro, R. J., and J. W. Lee. 2000. International data on educational attainment: Updates and implications. CID Working Paper 42, Harvard University.

Besley, T. 2006. *Principles agents? The political economy of good government.* Oxford: Oxford University Press.

Bienen, H., and N. Van de Walle. 1991. *Of time and power: Leadership duration in the modern world.* Stanford, CA: Stanford University Press.

Burnside, C., and D. Dollar. 2000. Aid, policies and growth. *American Economic Review* 90(4): 847–868.

Chauvet, L., and P. Collier. 2006. *Helping hand? Aid to failing states.* DIAL Working Paper 2006/14, Paris.

Chauvet, L., and P. Guillaumont. 2004. Aid and growth revisited: Policy, economic vulnerability, and political instability. In *Towards pro-poor policies*, ed. B. Tungoden (Proceedings of the World Bank Annual Conference on Development Economics), World Bank/Oxford University Press.

Collier, P. 2001. Consensus-building, knowledge and conditionality. In *Annual World Bank conference on development economics 2000*, ed. B. Pleskovic and N. Stern. Washington, DC: World Bank.

Collier, P., and D. Dollar. 2002. Aid allocation and poverty reduction. *European Economic Review* 46(8): 1475–1500.

Collier, P., and A. Hoeffler. 2004a. Greed and grievance in civil wars. *Oxford Economic Papers* 56(4): 663–695.

Collier, P., and A. Hoeffler. 2004b. Aid, policy and growth in post-conflict countries. *The European Economic Review* 48(5): 1125–1145.

Collier, P., and A. Hoeffler. 2006. Testing the neocon agenda: Democracy in resource-rich societies. Department of Economics, Oxford University, mimeo.

Collier, P., A. Hoeffler, and C. Pattillo. 2004. Aid and capital flight, mimeo, Centre for the Study of African Economies, Oxford University.

Devarajan. S., D. Dollar, and T. Holmgren. 2001. Aid and reform in Africa. Lessons from ten case studies. IBRD, Washington, DC: World Bank.

Dollar, D., and J. Svensson. 2000. What explains the success or failure of structural adjustment programs? *Economic Journal* 110(466): 894–917.

Fearon, J. D., and D. D. Laitin. 2004. Neotrusteeship and the problem of weak states. *International Security* 28(4): 5–43.

Gurr, T. R., B. Harff, and M. Marshall. 2003. *State failure task force report, phase 3: Findings*. Center for International Development and Conflict Management, University of Maryland.

Krasner, S. D. 2004. Sharing sovereignty: New institutions for collapsed and failing states. *International Security* 29(2): 85–120.

Lane, P. R., and A. Tornell. 1996. Power, growth and the voracity effect. *Journal of Economic Growth* 1: 213–241.

Little, I. M. D., R. N Cooper, W. M. Corden, and S. Rajapatirana. 1993. *Boom, crisis, and adjustment: The macroeconomic experience of developing countries*. New York: Oxford University Press/Washington: The World Bank.

Melhum, H., K. Moene, and R. Torvik. 2006. Institutions and the resource curse. *Economic Journal* 116(508): 1–20.

Miguel, E., S. Satyanath and E. Sergenti. 2004. Economic shocks and civil conflict: An instrumental variable approach. *Journal of Political Economy* 112(4): 725–753.

North, D. C. 1990. *Institutions, institutional change and economic performance*. New York: Cambridge University Press.

Rodrik, D. 1996. Understanding economic policy reform. *Journal of Economic Literature* 34(1): 9–41.

Ross, M. L. 2001. Does oil hinder democracy? *World Politics* 53(3): 325–361.

Sachs, J. 1994. Life in the economic emergency room. In *The political economy of policy reform*, ed. J. Williamson. Washington, D.C.: Institute for International Economics.

Sachs J., and A. Warner. 1995. *Natural resource abundance and economic growth*. NBER working paper 5398.

Sachs J., and A. Warner. 2001. The curse of natural resources. *European Economic Review* 45(4–6): 827–838.

Tavares, J. 2003. Does foreign aid corrupt? *Economics Letters* 79(1): 9–106.

Tornell, A., and P. R. Lane. 1999. The voracity effect. *American Economic Review* 89(1): 22–46.

World Bank (2004). *World development indicators*. CD-ROM.

Appendix 1—*Definition of the Country Policy and Institutional Assessment (CPIA)*.

A. Macroeconomic management and sustainability of reforms
  1. General macroeconomic performance
  2. Fiscal policy
  3. Management of external debt
  4. Macroeconomic management capacity
  5. Sustainability of structural reforms

B. Structural policies for sustainable and equitable growth
  1. Trade policy
  2. Foreign exchange regime
  3. Financial stability and depth
  4. Banking sector efficiency and resource mobilization
  5. Property rights and rule-based governance
  6. Competitive environment for the private sector
  7. Factor and product markets
  8. Environmental policies and regulations

C. Policies for social inclusion
  1. Poverty monitoring and analysis
  2. Pro-poor targeting and programs
  3. Safety nets

D. Public sector management
  1. Quality of budget and public investment process
  2. Efficiency and equity of revenue mobilization
  3. Efficiency and equity of public expenditures
  4. Accountability of the public service

Each of the 20 components of the CPIA is rated on a scale of 1–6.

Appendix 2—Data and Variables

Technical Assistance and Other Aid

Aid is defined as net disbursements of official development assistance, minus technical assistance. ODA disbursements and technical assistance data is from OECD, and was divided by the Gross National Product (GNP) from World Development Indicators of the World Bank (2004).

Instruments for Aid

Same Language as Donor i. Dummy taking the value of one if the donor country and the recipient country share a common language [from Collier, Hoeffler, and Pattillo (2004); source: CIA factbook (2003); http://www.cia.org/publications/factbook].

Same Religion as Donor i. Dummy variable taking the value of one if 30% or more of the population belong to one religious group in the donor as well as in the recipient country [from Collier, Hoeffler, and Pattillo (2004), source : Barrett (1982)].

Total Aid Budget of Donor i. Total net disbursements of ODA by donors *i*, in constant prices 2001 (source: OECD; http://www.oecd.org/).

CPIA

Country Policy and Institutional Assessment (World Bank). It has 20 equally weighted components, divided into four categories: (1) Macroeconomic management and sustainability of reforms; (2) Structural policies for sustainable and equitable growth; (3) Policies for social inclusion; (4) Public sector management.

Income Per Capita

Real gross domestic product *per capita* ($ in 1996 constant prices), Penn World Tables 6.1.

Postconflict Years 1 to 4

Dummy variable equals to one in years one to four after the ending year of a civil war (Collier and Hoeffler, 2004b).

Secondary Education

Barro and Lee (2000) dataset. Percentage of the population who completed secondary education (population aged 25 and above).

Population

World Development Indicators of the World Bank (2004).

Democracy

Polity IV score for democracy. Ranges from 0 to 10 (0 = low; 10 = high). Measures the general openness of political institutions. The 11-point Democracy scale is constructed additively. The operational indicator is derived from coding of different authority characteristics.

Years Spent in Office

Years the national leader had been in office. "0" indicates transition year, i.e., a year of change in leadership. Source: State Failure Task Force (Gurr, Harff, and Marshall, 2003) (http://globalpolicy.gmu.edu/pitf/) and Bienen and van de Walle (1991).

Resource Rents

Natural resource rents (% of GDP). Resource rents are calculated by summing the total value of rents for all extractive (i.e. nonagricultural) resources. These values are in current US$ and are then divided by GDP in current US$. Source: Collier and Hoeffler (2006).

State Fragility and Implications for Aid Allocation: An Empirical Analysis

DAVID CARMENT
YIAGADEESEN SAMY
STEWART PREST

> Fragility belongs to a class of properties, or qualities, philosophers call dispositional. A dispositional property differs from a static one in being lodged in potential rather than actuality. An antique china teacup is always fragile and it is also always white but its whiteness is bodied forth at every moment whereas its fragility depends on what will happen under certain future circumstances.[1]

Introduction

Despite its importance as part of the international policy discourse, the idea of state fragility remains an elusive concept, difficult to quantify and to evaluate. Definitions by different organizations identify 30 to 50 fragile states, many of which are experiencing or have experienced protracted conflict. Collectively, these states represent a significant portion of the world's population at risk of not meeting the UN Millennium Development Goals. Given the recent findings by Chauvet and Collier (2004) on the high cost (for themselves and their neighbors) of countries that fail, it is important to understand how states become fragile and why, if development assistance is to have a measurable and positive impact on fragile states.

Using CIFP's[2] fragility index for the period 1999–2005, this article provides an empirical analysis of state fragility using a parsimonious model based on the idea that a state needs to exhibit three fundamental properties of authority, legitimacy, and capacity (ALC) and that weaknesses in one or more of these dimensions will have an impact on the overall fragility of a particular country. Authority refers to the ability of the state to enact binding legislation over its population and to provide the latter with a stable and safe environment. Legitimacy refers to the extent to which the governing regime enjoys public loyalty and support for government legislation and policies along with international recognition of that support. Capacity refers to the power of a state to mobilize public resources for productive uses.

A second and related objective is to evaluate the relationship between ALC properties and aid allocation and aid effectiveness. We pursue this objective because we believe that proper fragile states analysis is crucial to proper aid allocation decisions. More specifically, not only must policymakers be sensitive to fragile states environments when making decisions on aid, aid allocation that focuses on authority and capacity considerations alone, may be counterproductive. Using our ALC framework, we show that a key factor in determining long-term state stability, namely legitimacy, is generally underemphasized in aid allocation decisions. Indeed, our results indicate that aid allocation to fragile states correlates strongly with measures of authority and capacity, and weakly with legitimacy.

Our argument is that policy relevance and analytical utility stem from a broader appreciation of the causes of fragility than that which would be encompassed by focusing on the immediate determinants of failure, such as challenges to state authority and violent conflict. We argue that fragility and failure are analytically distinct. While the two concepts share some properties, they differ in the degree to which they emphasize the roles of armed conflict and instability in contributing to state weakness. This distinction is especially pertinent to donor countries whose policy tools have been redesigned to extend beyond operational responses (e.g., the use of force and diplomacy to prevent or mitigate violence and instability) and are intended to shore up structural weaknesses and guard against further neglect. Implicitly in our conceptualization of state fragility, we recognize that states become fragile and fail for different reasons and that they are qualitatively different from one another, with unique problems that often require distinct policy responses.

The theoretical implication of our argument is simple. The bulk of research in this field has almost exclusively focused on defining the dependent variable in terms of failure and there is a heavy bias in selecting causal factors on the dependent variable, a point made clearer in the literature review below. As a result, research programs tend to conceptualize failure as an end state, associated with either or both armed conflict or low capacity. The policy implications of this research are also clear. By focusing on end states, and in particular those brought about by violent events or low capacity, our ability to anticipate

[2]Since 1997, the Country Indicators for Foreign Policy (CIFP) project (together with the Canadian government, private sector and nongovernmental organizations) has collected and analyzed statistical information on a range of issues related to the political, economic, social, and cultural environments of countries around the world (see www.carleton.ca/cifp).

earlier turning points and stages where a fragile state might or might not fail is held back. For example, Pakistan has been a fragile state for some time now, yet according to some research programs Pakistan is categorized as failing only after it experiences large-scale violence and instability.[3] Arguably, once a country like Pakistan enters into "failed state" status, it is simply too late to do anything but shore up existing institutions and structures, no matter how weak or corrupt they are, as a bulwark against further decline.[4] This "shoring up" of authority structures then results in a vicious cycle of further decline where both capacity and legitimacy are undermined and in turn authority is further challenged. The overall claim being made here is that misdiagnosis has the potential to lead to the misallocation of policy resources, which in turn can lead to ineffectiveness and further state decline.

The rest of the paper is organized as follows. Section two reviews the dominant research streams in order to place our work in context. In section three, we describe the methodology used to operationalize our ALC framework. In section four, we specify the model(s) tested, and discuss the empirical results pertaining to the determinants of fragility. Section five examines the issue of aid allocation and effectiveness in fragile states. We conclude with the policy implications of our analysis and directions for future research.

The Three Streams of Fragility Analysis

In this section, we identify how fragility is conceptualized in both policy discourse as well as empirical research. Recognizing that there are numerous research projects from which to draw, we assess three policy-inspired research streams that have contributed to the concept of fragility. Broadly speaking, they are development-oriented, conflict-oriented, and stability-oriented. It is the convergence of these three streams of thought that informs most discussions of state fragility today. As policymakers continue to grapple with the distinct, though interconnected, policy discussions related to these streams, some have begun to focus on the fact that many of these diverse problems are in fact geographically confined to the same limited number of states globally. The concept of fragility therefore represents an effort by policymakers to generate a conceptual framework sufficiently general to capture these disparate, and indeed somewhat contradictory, influences on foreign policy that focus on a finite number of states in specific regions of the world.[5]

Though rooted in much earlier discussions, the immediate antecedents to fragility analysis emerged out of global policy developments of the 1980s and 1990s. The first research stream, championed by the World Bank but echoed by DfID and the OECD among others, examines the reversals suffered by the states that attempted to implement structural adjustment and market-based reforms. The general experience of such states was

[3]Compare the disparate rankings for Pakistan in the 2005 and 2006 editions of the Fund for Peace Failed States Index and then consider our ranking in Carment et al. (2006). In 2005, FFP ranked Pakistan 34th on their list, while in 2006, Pakistan ranked 12th (Fund For Peace, 2005, 2006). While the FFP list may provide some information regarding the relative level of violence in the state, it is unclear what else could drive such a shift in rankings. Moreover, by focusing on failure as its dependent variable, such an index provides little basis for any sort of forward-leaning policymaking, as it provides more information about where crises are currently occurring than when and where they may emerge.

[4]Such an emphasis, exemplified in U.S. support for allies such as Pakistan in the global war on terror (GWOT), leads to potential distortions in both the selection of aid recipients, and the types of aid provided. Aid is provided to states with limited authority and capacity, regardless of the relative legitimacy of the regime resulting in deeply unpopular regimes heavily dependent on external aid that can be unstable over the long term. U.S. aid to Pakistan is predominately military aid.

[5]It should be noted that the identified research programs within these streams do not themselves always use the term fragility. We use the term fragility here as a concept that contains elements that encompass all three streams.

one marked by turbulence, retrenchment of state activities, and only limited, often highly unequal growth (Gore, 2000; Stiglitz, 1998). Some states proved particularly vulnerable to the negative aspects of structural adjustment, so much so that they earned particular designations among development agencies and researchers as difficult partners (OECD, 2001), difficult environments (Moreno-Torres and Anderson, 2004) and low-income countries under stress (LICUS) (World Bank, 2002). Regardless of the name, they were states in which development as usual appeared to be ineffective or even counterproductive. Aid agencies and international organizations began the search for alternative strategies to be employed in such situations; the Paris Declaration on Aid Effectiveness (OECD, 2005) is broadly representative of such efforts.[6]

A second research stream grew out of the conflict prevention and early warning agendas that dominated much of the 1990s. Beginning with Boutros Ghali's *Agenda for Peace* (1992), through the global trauma represented by Rwanda, Sierra Leone, and Somalia, to Haiti, to the Solomon Islands and East Timor, to the qualified successes of Bosnia and Kosovo, a great deal of discussion among policymakers and academics alike focused on the global shift from interstate to intrastate conflict that occurred through the latter part of the Cold War, and continued throughout the 1990s. Seminal contributions to this stream include the *Carnegie Commission on Preventing Deadly Conflict* (1997), the *Responsibility to Protect* (ICISS, 2001), and the *Human Security Report* (Human Security Centre, 2005). In this vein, the conflict prevention research by Baker and Ausink (1996), which sought to develop a predictive model of ethnic conflict, served as a precursor to their subsequent Fund for Peace State Failure Index. Rooted within the literature on violent intrastate conflict, the Fund for Peace Index promotes a definition of failure that focuses on a subset of related factors, including demographic pressure, movement of refugees, economic shocks, and other late-stage triggers and accelerants (Fund for Peace, 2007). With such a strong focus on effects, rather than causes, the result is a tool arguably more descriptive of states that have already substantially failed or are in imminent danger of doing so, rather than those with the potential to fail over the medium or long term.

Brecher and Wilkenfeld's International Crisis Behaviour project reflects a similar line of thinking, though employing interstate crisis rather than intrastate conflict or failure to operationalize its analysis (1989). The concept of crisis response figures prominently in wider discussions of fragility as well, though expanded beyond Brecher and Wilkenfeld's focus on interstate crisis to encompass states' ability to respond to economic crises, natural and humanitarian disasters, pandemics, and other exogenous shocks. For example, the United Kingdom's Prime Minister's Strategy Unit report on instability incorporates the idea of exogenous shocks in its fragile-states model, specifying the inability to respond effectively to these shocks as a defining feature of fragile states, and making increased response capacity a core goal of any international engagement (United Kingdom, 2005).

The third stream relates to the threats that weak and failed states represent to neighboring states and to the international system because of their tendency toward instability and loss of authority. Whether through direct sponsorship of terrorist activities, as in the

[6]The OECD defines fragile states as "countries where there is a lack of political commitment and insufficient capacity to develop and implement pro-poor policies" (Morcos, 2005). USAID uses the term fragile states to refer to a broad range of failing, failed, and recovering states, admitting that the distinction among them is not always clear in practice; they also identify states that are vulnerable from those that are in crisis, and report that about a third of the world's population now lives in areas that are fragile. The World Bank, on the other hand, focuses its attention on 30 low-income countries under stress (LICUS) that are characterized by a combination of weak governance, policies, and institutions, and ranked among the lowest on the Country Policy and Institutional Assessment (CPIA); the 30 LICUS countries comprise about half a billion inhabitants.

case of the Taliban's support for Osama bin Laden in Afghanistan, or simply the fact that such states constitute zones of impunity in which transnational criminal organizations, pirates, and internationally focused terrorists operate, or again because such states with their limited infrastructures are unable to cooperate with other countries in the containment of common threats to human health represented by avian flu, pollution, and environmental change, such states constitute gaps in the international order.

Obviously, such considerations were pushed to the forefront in the years following September 11, 2001. They were clearly present prior to that date, however. Berger's emphasis on early intervention in foreign conflicts as being in America's foreign policy interest provides an example of such thinking (2000, 29). More recently, the UN High Level Panel on Threats, Challenges, and Change provided a very broad assessment of the threats currently requiring collective international action. Virtually all of them relate in some way to what the report refers to as weak states (2004). It is, above all, the mutual vulnerability of weak and strong, or more precisely, the vulnerability of the strong to the dangers posed by the weak, that motivates such discussions.

In terms of both purpose and methodology, the State Failure Task Force, now known as the Political Instability Task Force (PITF), exists within this third research stream.[7] Using a primarily conflict-derived definition of state failure focusing on rebellion, genocide and politicide, adverse regime change, and revolutionary or ethnic wars, the PITF has attempted to identify a limited number of conflict predictive factors that could be used to inform security policy. The original task force's analysis identified three such factors: infant mortality, openness to trade, and the level of democracy (Esty et al., 1998). Subsequent studies by Goldstone et al. (2005) revised this model, disaggregating polity type into two discrete measures, executive recruitment and competitiveness of political participation, drawing on the work of Dahl (1971) in doing so. Goldstone and colleagues found that their model improved markedly when these disaggregated elements interacted with the level of factionalism encountered in the regime, as measured by the Polity dataset (Hewitt, Wilkenfeld, and Gurr, 2007).

According to the authors, the interactive effects of these factors, when combined with variables capturing the level of conflict in contiguous states, and the logged and normalized measure of infant mortality (considered to be a proxy for standard of living), create a highly predictive model, capable of out-performing much more sophisticated models using a variety of other variables.[8] With funding from the United States Agency for International

[7]According to the original State Failure Task Force a failed state is one that is "utterly incapable of sustaining itself as a member of the international community" (Esty et al., 1995, 1). Narrowly defined however, *"state failures consist of instances in which central state authority collapses for several years"* (Esty et al., 1995, 1); emphasis. However, since fewer than 20 such episodes have occurred during the last 40 years, it is difficult to conduct statistical analysis. Therefore, the initial task force broadened the concept of state failure to include a wider range of civil conflicts, political crises, and massive violations of human rights that are typically associated with state breakdown. In line with such a broad definition, the task force isolated four kinds of state failure: (1) revolutionary wars, (2) ethnic wars, (3) mass killings, and (4) adverse or disruptive regime change.

[8]To be sure, there are a variety of other elements posited to have an effect on conflict, state stability, or both. For instance, Cramer (2005) summarizes the relationship between inequality and conflict, concluding that most authors think there is a relationship, but virtually no consensus regarding its nature. To cite a limited number of examples, Muller (1985), invoking Aristotle and de Tocqueville and presenting empirical evidence in support, suggests that individual inequality is positively correlated with political violence. Nairn (1977) finds a link between class-based inequality and violence, while Holsti (1995) and Gurr (1993) highlight the effects of intercommunal inequality. Urdal (2004) finds that the "youth bulge" has a significant effect on the incidence of conflict in states. A number of writers, including Collier et al. (2003) and Hegre et al. (2001), highlight the significant relationship between the prior experience of instability and conflict, and their likely reoccurrence in the future.

Development (USAID), Marshall and Goldstone (2007) have since integrated these and other insights into a new Fragile States Matrix. The matrix now measures state effectiveness, as well as legitimacy across four dimensions: political, economic, social, and security.[9]

In brief, individually these three streams constitute an important part of understanding fragility. Yet they also result in policy approaches that are only partially complementary. They define fragile and failed states differently, generate different lists of at-risk states, and prescribe different policy approaches. For example, the FFP Failed States Index exists in that same tradition as the PITF but is distinct with its primary focus on emergent or existing conflict rather than political instability. Failed and fragile state research agendas most concerned with stability and conflict tend to encourage policies that strengthen authority structures. Unlike the FFP or the PITF, the World Bank's research on Low Income Countries Under Stress (LICUS) places capacity, the ability of the state to deliver core public goods and services unassociated with anti-insurgent activities, front and center. Having identified how fragility is conceptualized in both policy discourse as well as empirical research, we now turn to our alternative framework.

Operationalizing Fragility: ALC Defined

The three pillars of our framework are authority, legitimacy, and capacity. Each is considered in turn. Weber identified authority and legitimacy to be two related, though conceptually distinct, elements of power ([1922] 1968, quoted in Uphoff, 1989). Authority, for Weber, is a measure of the probability that a given group of individuals will obey a given command. Legitimacy is defined as "a conviction on the part of persons subject to authority that it is right and proper and that they have some obligation to obey, regardless of the basis on which this belief rests" (Uphoff, 1989, 301). Writing in 1973, Hurwitz synthesized a number of different themes present in the literature on stability at that point, as researchers during the previous decade had attempted to operationalize the concept in a number of different ways. One of these approaches, described by Hurwitz as "a multifaceted societal attribute," seems to presage many of the discussions surrounding the concepts of fragility today. Citing studies by a variety of pioneers in the field, including Lijphart (1968) and Eckstein (1966), who focus on the presence of various criteria such as persistence, legitimacy, effectiveness, and authenticity, Hurwitz introduces a conceptualization of stability reminiscent of modern discussions of fragility (Eckstein, 1966, 11–32).

Our approach is a synthesis of these theoretical foundations as well as the contemporary research streams described above. Much of that discussion regarding the role of conflict informs the authority pillar within our ALC framework. Other components of authority involve state control of borders and the presence of widespread illicit activities within a country. For example, Lake and Rothchild (1998) argue that the presence of conflict on a state's borders is a significant determinant of conflict within the state, positing

[9] In their systems dynamics model of state stability, Choucri et al. (2006) define state stability such that "a state is stable to the extent that its resilience (capabilities) is greater than the load (or pressures) exerted upon it" (Choucri et al., 2006, 2). Choucri et al. operationalize "loads" as the relative pressures exerted by insurgent activities, echoing the PITF. Some authors focus on the relationship between political systems and relative stability, focusing on the link between state legitimacy and stability. Hegre et al. (2001) are representative of a common school of thought, also supported by the Political Instability Task Force, that holds that inconsistent regimes experience greater instability, as measured by the relative incidence of civil war, than either democratic or autocratic regimes. Notwithstanding policy statements to the effect that increased democracy will necessarily decrease the risk of both intrastate and interstate conflict globally, there is little evidence supporting such a linear relationship; there is a growing consensus around the more nuanced "inverted-U" hypothesis articulated by Hegre et al. (2001) at the intrastate level, and Mansfield and Snyder (1995) at the interstate level.

that diffusion and contagion effects are responsible for much of the violence seen in the post–Cold War era. Several research programs argue that contract enforcement becomes salient in the absence of a state with a functional regulatory apparatus. Under this formulation, when the latter is present, resource availability and other determinants of rent-seeking behavior become relatively insignificant to the stability of the state. Dunning (2005) emphasizes the critical role played by elites, and their payoff structures, in determining whether resources become the objects of destabilizing political competition.

On the issue of legitimacy, Alao (1999) and Herbst (1996/7) find a correlation between a state's colonial legacy and its subsequent stability. Illegitimate postcolonial states are inherently unstable and this loss of legitimacy further hampers a state's ability to control conflict within and beyond its borders. Others consider politicized ethnicity a particularly potent force used to mobilize populations and thereby reduce the legitimacy of the broader state identity. Such authors search for the particular facets of intercommunal relations most likely to provoke distrust, violence and, by extension, loss of authority and legitimacy. In his Minorities at Risk project, Gurr (1993) codifies the specific experiences of identifiable mobilized minority groups around the world. On the basis of this systematic and expansive body of evidence, Gurr argues that economically, politically, or socially repressive actions by majoritarian regimes, or in a limited number of cases, repressive actions by a privileged minority against a disadvantaged majority, are the most salient factors in predicting intercommunal instability and conflict.[10]

With respect to the state capacity pillar, Fearon and Laitin (2003) argue that state stability is largely determined by factors related to the viability of an insurgency, which in turn is heavily dependent on state capacity. Other factors include economic growth and inequality as indicators that favor recruitment; low per capita income, which acts as a proxy for governmental ability to repress potential insurgents; and large population and rugged terrain, both factors associated with more viable insurgency.[11] Gupta (1990) draws on development economics theory to generate a growth model that includes political stability as an endogenous element. He finds that, among other things, political instability is associated with reduced economic growth. Alberto Alesina et al. (1996) attempt to clarify the direction of causality in the relationship between growth and stability. They find that political instability tends to reduce growth rates, though low growth rates result in

[10]A variety of other studies contribute to our understanding of legitimacy as well. For instance, Fein (1995) suggests a link between the level of political and civil rights in a state and the likelihood of political stability, finding an inverted "U" relationship similar to that found for regime type. Luckham et al. (2001) posit a link between the level of poverty of sub-Saharan Africa and the endemic nature of interstate conflict experienced there, suggesting that the region is particularly vulnerable to the development of conflict-sustaining political economies. Stedman (1996) and Van de Walle (2001) find a link between negative economic growth and instability consistent with Gurr's deprivation theory (1970).

[11]A separate line of argument maintains that fragile states are the product of largely structural factors, including geography, access to the sea (or other international trade routes), the presence of sufficient population, and economic capacity to sustain centralized government. Diamond (1997), Collier (2007), and Collier and Hoeffler (1998) have contributed to this school of thought, although each identifies different processes contributing to the current state of radical inequality found in the world. While Diamond emphasizes the role that physical geography has played on the accumulation of human capital in different regions of the world over the very long term, Collier and Hoeffler (1998) focus on more proximate approach, suggesting that the so-called resource-curse plays an important role in the relative vulnerability of states to violent conflict. According to this argument, countries that are highly resource dependent are at an increased risk of conflict, as such resources can be used to fuel either government intransigence or rebellion. This argument has subsequently been refined to focus particularly on lootable resources, including drugs, secondary diamonds, timber, rubber, and a limited number of precious metals, which tend to be associated with elevated levels of conflict.

reduced stability only in specific circumstances; in other words, lapses in state authority can have a negative effect on state capacity, while limited state capacity does not necessarily impair authority. Others, from early work by Gurr (1970) on relative deprivation to recent macroeconomic studies by Feng (1996) on the effects of economic growth on political stability, maintain that economic growth does indeed have a significant effect on stability, with poor or negative growth associated with an elevated risk for political violence and regime change. Focusing on economic aspects of stability, Ramey and Ramey (1995) find that countries with greater short-term volatility exhibit lower growth over the long term. Given other studies linking poor economic growth to increased political instability, one would expect a commensurate link between economic and political volatility.

International trade linkages are also important measures of capacity. The effects of trade and globalization remain ambiguous; Barbieri and Schneider (1999) find that trade has no pacifying effect on interstate conflict, though Russett and Oneal (1999) dispute those findings and reiterate support for the liberal thesis first articulated by Kant ([1795] 2003). Goldstone et al. (2005) maintain that openness to trade is a primary determinant of state stability. Reuveny (2000) argues that conflict over the distribution of the gains from trade may be a more interesting question than the effects of trade on interstate conflict, suggesting that future studies of the relationship between trade and conflict should model the relationship dynamically in order to capture the two variables' reciprocal causality; moreover, he argues that such studies should be careful to disaggregate the interests of importers, exporters, and the government, and to incorporate considerations of the types of goods and services being traded into the analysis.

Charles Tilly (1978; 2003) develops a similar argument, suggesting that violent group interaction normally results from a complex series of social activities necessary to group mobilization. He too suggests that levels of democracy and state capacity play a role in the relative level of violence experienced, though his goal is to explain the relative level of violence, rather than the relative stability of various systems of government; the two projects are similar, though not identical in nature. Other things being equal, Tilly maintains that low-capacity regimes experience more violence than their high capacity counterparts.

Finally, Robert Rotberg (2003), building on the work of others, combines elements of capacity and authority in his definition. Rotberg characterizes failed states as being marked by an inability to provide basic political goods, including human security and security of the state, dispute resolution and norm regulation, essential political freedoms, and economic opportunity to most, if not all, of the population. For Rotberg, a fragile state is one that cannot reliably control its own territory, as in the case of Sri Lanka or Colombia, or one that cannot provide essential public goods such as basic education and health care, infrastructure, a predictable market through contract enforcement and other elements of capitalism.

In operationalizing fragility, our study adapts and operationalizes the two Weberian concepts of legitimacy and authority, and combines them with the concept of capacity, the ability of a state to mobilize resources for the provision of public goods, as highlighted by Tilly (2003), Fearon and Laitin (2003), Rotberg (2003), and many others as being integral to the relative fragility of a state. Thus, we arrive at the following threefold formulation: *fragile states lack the functional authority to provide basic security within their borders, the institutional capacity to provide basic social needs for their populations, and the political legitimacy to effectively represent their citizens at home and abroad.*[12]

Authority refers to the ability of a functional state to enforce binding legislation over its population, and exercise the coercive force over its national territory necessary to provide

[12]For a complete exposition of the 80 indicators that are included within each of the six clusters of our ALC framework see Carment et al. (2006).

a stable and secure environment to its citizens and communities. It is a measure of a state's ability to project force. Security is a necessary prerequisite of the realization of public, private, and civil society interests. States lacking in authority may be unable to exercise control over the full extent of their legal territory; such states will likely have difficulty responding effectively to threats, whether internal or external. In some areas, nonstate actors such as rebel militias or criminal organizations may exercise de facto authority, while in others the rule of law may be completely absent. Border control may be intermittent or nonexistent, enabling illicit flows of people and goods. State response to foreign incursions may be weak and ineffective. Other potential problems include the inability to enforce government policy, combat corruption and criminality, effectively mobilize the resources of the state toward the ends requested and required by government, regulate private markets, or guarantee contracts.

Legitimacy refers to the ability of a state to command public loyalty to the governing regime, and to generate domestic support for government legislation and policy. Such support must be created through a voluntary and reciprocal arrangement of effective governance and citizenship founded upon broadly accepted principles of government selection and succession that are recognized both locally and internationally. States in which the ruling regime lacks either broad and voluntary domestic support or widespread international recognition suffer a lack of legitimacy. Such states face significant difficulties in maintaining peaceful relations between and among various communities within the state; any security found within the state is likely the result of coercion rather than popular consent. As a result, such states are inherently vulnerable to internal upheaval, and are likely to remain fragile so long as legitimacy remains wanting.

Capacity refers to the power of a state to mobilize and employ resources toward productive ends. States with a satisfactory level of capacity display a basic competence in political and economic management and administration, with their governments capable of regulating domestic affairs and conducting international transactions. They also possess the basic infrastructure required of a modern state, including functional transportation and communication networks.[13] States lacking in capacity tend to be heavily dependent upon civil society and the international community for the provision of vital public goods and services, and may prove unable to respond effectively to sudden shocks such as natural disasters, epidemics, food shortages, or refugee flows.

Our model is constructed through the selection of structural indicators that represent and collectively encompass the three pillars of authority, legitimacy, and capacity. While some indicators are selected on the basis of their relevance specifically to the risk of violent conflict and instability, fragility assessment requires a much broader range of measures. The predominant foci of research on state fragility are issues of governance, security, economic development, human development, and the environment, all of which are captured within the ALC framework.

Table 1A in the Appendix provides a list of the top 40 fragile states based on data for the period 1999 to 2005, with Burundi being the most fragile state. The table provides data for the overall fragility score, scores along the ALC components, and scores for the different

[13]The development of political capacity, legitimacy, and authority, all essential features of state-building, is not a linear process. This is especially relevant to explaining state failure, since changing environmental conditions can reverse (in very short periods of time, e.g., months and years) some or all of these essential features. For example, changes in system structure can reverse state-building in at least two non–mutually exclusive ways: through the creation of highly dependent weak states (compounded by the subsequent withdrawal of powerful patron states) on the one hand; and through processes of economic development, and the strengthening of international norms of self-determination, on the other.

indicator clusters. Gender acts as a cross-cutting theme. What is striking in examining the most fragile states is that though they appear in more than one category they rarely rank high on all three measures of A, L, or C, an indication that fragility manifests itself in different forms that require different forms of intervention and engagement. When broken down in terms of the six indicator clusters (e.g., for the top 20 fragile states), none appears on all six and only a few appear on four, again an indication that fragile states face different challenges that a one-size-fits-all approach is unlikely to resolve.

A Model of Fragility: Empirical Results

There are no rigorous empirical studies of the determinants of state fragility, and very few papers have examined aid allocation and effectiveness for fragile states. We undertake both tasks in this section and the next one. In this section, we validate the CIFP fragility index constructed on the basis of the theoretical literature reviewed above. The index, using approximately 80 indicators to capture various facets of authority, legitimacy, and capacity, provides a relative ranking of all countries. We compare this index with the primary determinants of fragility identified in the literature for a cross-section of countries over the period 1999–2005. Having reviewed the existing literature on state fragility and having discussed conceptions of fragility and failure, we estimate a model of fragility, where the latter is determined by a range of economic, political and social factors. More precisely, drawing on the work of the PITF and literature relating economic growth and stability, we adopt the following baseline model:

$$\text{fragility}_i = \beta_0 + \beta_1 \text{income}_i + \beta_2 \text{growth}_i + \beta_3 \text{demo}_i + \beta_4 \text{trade}_i + \beta_5 \text{inmo}_i + \varepsilon_i \quad (1)$$

where the different variables are operationalized as follows. Fragility is the CIFP fragility index described in the previous section, with only countries with a fragility score of 4 and above being considered.[14] Excluded countries are essentially OECD countries as well as a few high-income countries such as Singapore and South Korea. The independent variables include: Income, which is captured by the logarithm of gross domestic product per capita in purchasing power parity (PPP); Growth refers to the growth of gross domestic product per capita; Demo refers to the level of democracy from the Polity IV dataset and it varies from strongly autocratic (−10) to strongly democratic (+10); Trade refers to trade openness, namely the ratio of the sum of exports and imports to gross domestic product; Inmo refers to the infant mortality rate. Subscript i refers to countries and ε_i is the normal disturbance or error term with the usual properties. Our model, therefore, considers the most common factors identified in the literature and then verifies whether other factors such as income inequality, human rights, ethnic risk and ethnic diversity are also important determinants of state fragility by augmenting the baseline model. Table 3A in the Appendix provides a correlation matrix of all the variables used in the regression analysis. All the regressions include regional dummies to account for geography as an additional factor.

Table 1 provides summary statistics for the variables used in the empirical analysis, arranged in alphabetical order. We have data for more than 100 countries for most of the variables. The average value of the fragility index is within the range of countries performing around the median value. Except for GDP per capita, the mean and median values are not too different and the standard deviations are not too large for the other variables; in the case

[14]Within the CIFP Index, states are scored from 1 to 9 on each indicator included, with 9 being the most fragile (see Table 2A in the Appendix for the fragility index scoring scale). Relevant scores are then averaged to provide overall scores for A, L, and C, as well as for overall fragility. For a full description see Carment et al. (2006).

TABLE 1 Summary statistics

Variable Name	Description	Number of observations	Mean	Median	Standard deviation
ETHRISK*	risk of rebellion	98	4.84	3.77	3.86
FRG	fragility index	156	5.89	5.80	1.06
GDPPC	gdp per capita	134	5031	3867	4552
GINI	gini coefficient	100	42.33	40.58	9.94
GRW	economic growth	148	3.46	3.56	3.11
HDI	human development index	148	0.65	0.72	0.16
HREM	human rights—empowerment	154	5.32	6.00	3.07
INMO	infant mortality rate	155	51.47	36.80	39.20
LDEM	level of democracy	129	1.77	4.00	6.40
LIT	literacy rate	107	77.73	84.00	20.34
POPDIVET*	ethnic diversity	128	0.39	0.42	0.26
POPDIVREL	religious diversity	134	0.39	0.38	0.23
TRAD	trade openness	141	83.18	75.38	37.71

*Single measures.

of GDP per capita, the log transform of the variable will be considered in the regression analysis in order to normalize the distribution.

Table 2 below shows the results when the benchmark Equation 1, and different specifications based on the latter, are estimated using OLS for all countries with a fragility index of 4 and above. The level of development of a country, measured by the logarithm of GDP per capita, is highly significant and with the expected sign in column (1), which is the benchmark Equation 1; poorer countries tend to be more fragile than richer countries on average. When the level of development is replaced by a broader measure of development, HDI, in column (6), the same result is obtained; different specifications did not change the result either as seen in the other columns of Table 2. The other variables, namely growth, the level of democracy, trade openness, and infant mortality are also significant and with the expected signs. Over the estimated period, countries that grew faster tended to be less fragile, democratic countries displayed less fragility, countries that are more open to trade tended to be less fragile, and countries with higher infant mortality rates tended to be more fragile.

The possibility that the relationships were nonlinear was also investigated by adding square terms for the level of democracy, trade openness, and the human rights empowerment index. This relationship was confirmed in the case of the level of democracy only but not in the case of trade or human rights. More precisely, the inverted "U" relationship that several authors have uncovered between conflict and regime type is confirmed in the case of our fragility index as well. In column (4), when the human rights empowerment index is considered, it is highly significant with a negative sign when included on its own, indicating that countries with better human rights records tend to be less fragile. Finally, both variables for ethnic risk and ethnic diversity are significant and positively correlated with fragility (even though we have fewer observations in those cases). Income inequality (proxied by Gini coefficients) was not a significant factor when added to the benchmark equation (results not shown here), implying that the level of development, rather than income inequality, is important in explaining fragility. The overall fit of the estimated equations is greater than 80% in all cases, and the dummy variable for Africa was significant in most of the

TABLE 2 Determinants of fragility (dependent variable: fragility index, 4 and above)

Explanatory variables	(1)	(2)	(3)	(4)	(5)	(6)	(7)	(8)
Constant	9.81** (13.30)	9.44** (12.35)	9.80** (12.56)	9.85** (15.67)	9.69** (15.03)	9.39** (13.27)	9.89** (12.66)	9.37** (12.64)
log(GDPPC)	−0.47** (−6.06)	−0.40** (−4.77)	−0.47** (−6.00)	−0.43** (−6.18)	−0.42** (−5.98)	—	−0.51** (−6.41)	−0.43** (−5.33)
Growth	−0.05** (−3.47)	−0.05** (−3.24)	−0.05** (−3.37)	−0.05** (−4.42)	−0.05** (−4.27)	−0.05** (−2.67)	−0.06** (−3.71)	−0.06** (−3.25)
Demo	−0.04** (−4.67)	−0.03** (−4.21)	−0.04** (−4.60)	—	—	−0.03** (−3.10)	−0.03** (−3.99)	−0.04** (−4.38)
Demo*Demo	—	−0.01** (−2.06)	—	—	—	−0.01** (−3.20)	—	—
Trade	−0.01** (−3.45)	−0.01** (−3.69)	−0.01 (−0.88)	−0.01** (−3.90)	−0.01** (−3.87)	−0.01** (−2.95)	−0.01** (−2.20)	−0.01** (−3.58)
Inmo	0.01** (3.33)	0.01** (3.28)	0.01** (3.26)	0.01** (4.45)	0.01** (4.38)	0.01 (0.35)	0.01** (2.12)	0.01** (2.78)
Trade*Trade	—	—	−0.01 (−0.03)	—	—	—	—	—
Hrem	—	—	—	−0.10** (−6.48)	−0.05 (−0.96)	—	—	—
Hrem*Hrem	—	—	—	—	−0.01 (−1.19)	—	—	—
HDI	—	—	—	—	—	−4.36** (−5.26)	—	—
Ethrisk	—	—	—	—	—	—	0.05** (3.55)	—
Ethdiv	—	—	—	—	—	—	—	0.46** (3.05)
Afr	0.29** (2.29)	0.29** (2.24)	0.29** (2.24)	0.34** (2.93)	0.36** (3.01)	−0.03 (−0.30)	0.65** (5.00)	0.39** (2.66)
Lata	0.17 (1.33)	0.18** (1.44)	0.17 (1.33)	0.25** (2.20)	0.28** (2.28)	0.10 (0.73)	0.25* (1.81)	0.15 (1.13)
Mena	−0.01 (−0.04)	−0.05 (−0.32)	−0.01 (−0.05)	−0.09 (−0.73)	−0.08 (−0.64)	−0.22 (−1.37)	0.10 (0.85)	−0.02 (−0.13)
N	116	116	116	129	129	118	89	101
Adj. *R*-squared	0.83	0.83	0.83	0.86	0.86	0.83	0.86	0.85

Except where indicated otherwise, the figures in parentheses are the *t*-values. Standard errors are White-robust.

*$p < 0.1$.

**$p < 0.05$.

specifications (not surprisingly, given that many of the fragile countries in our sample tend to be from that region—see Table 1A in the Appendix).

Restricting the sample of countries to those with a fragility score of 6 and above (countries performing more poorly relative to others) reduces the overall fit of the equations as seen in Table 3. However, except for infant mortality rates and trade openness, the other independent variables remained significant and with the same signs. The results are thus quite robust; the level of development, economic growth, the level of democracy, ethnic risk, and ethnic diversity are all significant determinants of fragility. The fact that trade openness is no longer significant is not a surprise since most fragile countries tend to trade less in general.

Regressions reported in Tables 2 and 3 (and the causal inferences made from them) are often affected by omitted variables, reverse causality (endogeneity), or measurement errors.[15] Since we are using existing data at our disposal, and we have included as many explanatory variables as possible (based on the literature review), we are mitigating against measurement errors and omitted variables. Reverse causality remains a possibility; for example, state fragility is likely to influence a country's level of development, its growth rate, the extent to which it trades with the rest of the world, and even its political regime type. In order to make sure that our results are not affected by reverse causality, lagged values (by 5 years) of our independent variables were used in the different regressions and the results obtained were similar to those in Tables 2 and 3. Given that reverse causality remains a possibility (leading to biased OLS estimates), the standard method to deal with this problem is instrumental variable (IV) estimation. Instruments must fulfill two conditions. First, they must be correlated with the independent variables; second, they must not be correlated with the error term. Five-year lagged values of the independent variables were used as instruments (when available) since there is no correlation between them and the error term. IV estimation showed that the level of development remained a significant factor in explaining fragility while the other independent variables were no longer significant. In order to check the quality of instruments, residuals from Equation 1 were regressed on the instrumental variables, and none of the instruments were found to be significant.[16] Having confirmed the utility of the CIFP fragility index as a dependent variable, we now turn to the question of how the index can improve our understanding of aid allocation and effectiveness in fragile states.

Aid Policy and Fragility

Aid allocation and aid effectiveness remain controversial issues.[17] Early studies, including Papanek (1973) and Gupta and Islam (1983) found strong evidence that foreign aid flows are positively related with higher growth rates in recipient countries. This was followed by numerous other studies (for example by Mosley [1980] and Boone [1996] which found that aid did not contribute to growth. The Mosley (1980) study is quite important since it is one of the very first to include lagged values of aid as well as consider the possible endogeneity of aid using two stage least squares. Mosley (1986) also found that while

[15]All of these terms lead to the same econometric problem, namely that the independent variables are contemporaneously correlated with the error term.

[16]We have also tested for endogeneity in our analysis in order to make sure that our estimates are not biased.

[17]Aid allocation research answers the questions of who receives aid and why. Aid effectiveness research means large cross-country studies that have looked at the macroeconomic impact of aid on growth. The impact of aid has also been examined at the micro (project) level and in single-country case studies, not within the scope of this study.

TABLE 3 Determinants of fragility (dependent variable: fragility index, 6 and above)

Explanatory Variables	(1)	(2)	(3)	(4)	(5)	(6)	(7)	(8)
Constant	10.14** (9.89)	10.08** (9.88)	10.71** (8.80)	10.18** (12.80)	10.16** (12.46)	8.98** (10.31)	9.46** (7.82)	8.93** (7.29)
Log(GDPPC)	−0.44** (−3.59)	−0.42** (−3.39)	−0.48** (−3.67)	−0.41** (−4.24)	−0.41** (−4.19)	—	−0.41** (−2.73)	−0.30** (−2.01)
Growth	−0.05** (−1.95)	−0.05** (−1.95)	−0.05** (−2.06)	−0.04** (−2.22)	−0.04** (−2.17)	−0.05** (−2.06)	−0.03** (−1.39)	−0.08** (−3.07)
Demo	−0.05** (−4.26)	−0.05** (−3.92)	−0.05** (−3.79)	—	—	−0.03** (−3.00)	−0.05** (−3.65)	−0.04** (−3.44)
Demo*Demo	—	−0.01 (−1.27)	—	—	—	−0.01** (−3.20)	—	—
Trade	−0.01 (−1.35)	−0.01 (−0.93)	−0.01 (−1.17)	−0.01 (−0.53)	−0.01 (−0.52)	−0.01 (−0.74)	−0.01 (−0.44)	−0.01 (−1.53)
Inmo	0.01 (1.49)	0.01 (1.33)	0.01 (1.14)	0.01** (2.58)	0.01** (2.49)	−0.01 (−0.23)	0.01 (1.03)	0.01* (1.74)
Trade*Trade	—	—	0.01 (0.88)	—	—	—	—	—
Hrem	—	—	—	−0.13** (−7.20)	−0.12 (−1.56)	—	—	—
Hrem*Hrem	—	—	—	—	−0.01 (−0.18)	—	—	—
HDI	—	—	—	—	—	−3.19** (−2.73)	—	—
Ethrisk	—	—	—	—	—	—	0.05** (2.00)	—
Ethdiv	—	—	—	—	—	—	—	0.36* (1.91)
Afr	0.07 (0.42)	0.04 (0.22)	0.07 (0.42)	0.18 (1.34)	0.18 (1.34)	0.01 (0.04)	0.43** (2.55)	0.39** (2.66)
Lata	0.24 (1.06)	0.22 (0.98)	0.21 (0.92)	0.52** (2.84)	0.52** (2.81)	0.24 (1.01)	0.26 (1.13)	0.15 (1.13)
Mena	−0.06 (−0.23)	−0.18 (−0.66)	−0.04 (−0.15)	−0.38 (−1.77)	−0.38 (−1.68)	−0.21 (−0.75)	−0.20 (−0.69)	−0.02 (−0.13)
N	54	54	54	56	56	55	37	41
Adj. *R*-squared	0.59	0.59	0.58	0.73	0.73	0.56	0.72	0.64

Except where indicated otherwise, the figures in parentheses are the *t*-values. Standard errors are White-robust.

$^{*}p < 0.1$

$^{**}p < 0.05$

projects seemed to yield expected results, the macro evidence was not very positive, hence the so-called micro-macro paradox. Boone (1996) found that aid led to an increase in government consumption but did not increase investment or reduce poverty. Such studies provided support to one of the fiercest critics of foreign aid, Bauer (1981), who argued that just injecting money when the right incentives and institutions were missing would not force countries to grow. According to him, aid is a misallocation of resources because it is based on donor (as opposed to recipient) priorities, and tends to encourage corrupt or weak governments to remain in power.

Turning now to the relationship between aid flows and fragile states, Burnside and Dollar's seminal (and controversial) contribution (1997; 2000) showed that aid works in a good policy environment but not a bad one, such as would be found in a fragile state. Despite claims by Easterly (2003) that the Burnside-Dollar result is not robust when new data or alternative definitions are considered, the idea that aid works conditional on the policy environment continues to receive broad support.[18] More recently, Collier and Chauvet (2005) estimated that the cost of a country falling into LICUS status is US $80 billion on average, with most of the cost being shouldered by neighboring countries. If one takes the Burnside-Dollar argument seriously, as well as the fact that neglecting fragile states may in fact worsen poverty (and lead to further weakening), this implies that policymakers and analysts face a cruel dilemma. Should they be sensitive to fragile environments or not when making decisions on aid allocation? In answering this question, McGillivray (2007) examined the links between aid and economic growth in fragile states, classifying the latter as countries in the bottom quintile of the World Bank's Country Policy and Institutional Assessment (CPIA). A number of interesting findings are reported in McGillivray's study, namely that aid has a positive impact on growth in fragile states (but not as much as in nonfragile ones), and that fragile states have a lower absorptive capacity and are mostly under-aided, but the most fragile ones are over-aided.

Our own calculations based on OECD-DAC statistics show that fragile states are indeed under-aided relative to the overall sample of aid recipients. Table 4 shows how the pattern of aid differs from fragile states and all aid recipients, based on OECD-DAC statistics. When measured in terms of aid per capita, fragile states are underfunded relative to the overall sample and the volatility of aid flows has also increased over time. Furthermore, aid as a percentage of gross national income has not changed significantly (except for the period 1969–1978), indicating that fragile countries continue to be aid dependent. A closer look at individual countries (not reported here) shows that indeed some countries are over-aided (aid as a percentage of GNI) relative to others. For all countries in our sample (fragility scores of four and above), there is a positive relationship between aid as a percentage of income and the fragility index as seen in Figure 1.

The graph below does not tell us whether aid allocation is determined primarily by authority, legitimacy, or capacity of a state—the three pillars of state fragility. Our basic hypothesis is that aid allocation is determined by authority and capacity issues, whereas a key factor in determining long-term stability, namely legitimacy, is generally ignored in the policy decision-making process. This is confirmed in Table 5 below for different samples of fragile states, starting with all countries (fragility scores of 4 and above) and for smaller

[18]More recent studies (for example, Hansen and Tarp [2000] and Clemens, Radelet, and Bhavnani [2004] are more positive and argue that aid works, regardless of the policy environment, effectively nullifying the micro-macro paradox of aid. Other than the lack of an impact on growth, critics of aid have also referred to fungibility, lack of donor coordination, too much tying of aid, lack of absorptive capacity, and the failure of conditionality to buy reforms, among others.

TABLE 4 Aid allocation to fragile states

Variable recipients	1969–1978	1979–1988	1989–1993	1994–1998	1999–2003
All aid recipients					
Aid per capita (US $)					
-average	22.4	51.5	56.7	106.4	80.8
-standard deviation	30.4	64.9	70.7	518.9	202.5
Aid (%GNI)					
-average	5.9	8.9	10.6	10.7	8.8
-standard deviation	6.8	10.5	14.7	16.7	11.9
Top 40 fragile states					
Aid Per Capita (US $)					
-average	11.4	29.0	37.1	42.5	35.5
-standard deviation	10.3	21.6	23.8	38.6	46.1
Aid (%GNI)					
-average	6.5	11.3	15.7	16.0	13.8
-standard deviation	5.0	11.3	13.1	12.1	10.5
Top 20 fragile states					
Aid per capita (US $)					
-average	9.9	29.5	37.7	49.5	41.0
-standard deviation	7.1	19.7	22.8	48.8	62.6
Aid (%GNI)					
-average	6.1	12.1	16.0	17.6	15.9
-standard deviation	5.1	14.5	14.0	14.2	11.8

Source: Authors' calculations based on OECD-DAC aid statistics.

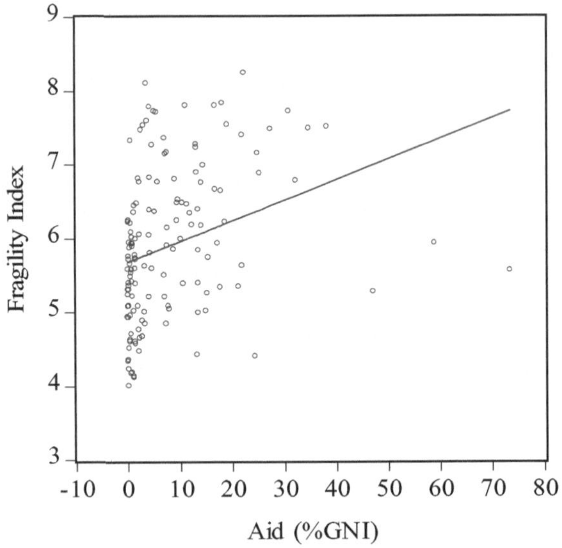

FIGURE 1 Aid and fragility (all countries).

TABLE 5 Determinants of aid allocation (dependent variable: aid/GNI)

Explanatory variables	All	5>	6>
Constant	−2.64 (−0.51)	1.42 (0.16)	−29.19** (−2.64)
Authority	−2.60** (−3.10)	−3.28** (−2.95)	−0.63 (−0.73)
Legitimacy	−1.54 (−1.53)	−1.56 (−1.19)	−1.55 (−1.03)
Capacity	5.70** (6.29)	5.74** (6.14)	7.43** (6.04)
N	145	117	60
Adj. *R*-squared	0.33	0.31	0.36

Except where indicated otherwise, the figures in parentheses are the *t*-values. Standard errors are White-robust.
*$p < 0.1$.
**$p < 0.05$.

subsamples (more fragile countries).[19] In fact, as we restrict the sample to the most fragile states, only capacity remains a significant determinant of aid allocation. An improvement in capacity is associated with less aid, while an improvement in authority is associated with more aid. Another way to interpret this result is that regardless of whether a state is strong or weak, it will receive aid independent of whether its legitimacy is weak or strong, and that the underfunding reported by McGillivray (2007) and confirmed by the numbers in Table 4, is not tied to questions of legitimacy.

A related point to keep in mind as far as aid and fragile states are concerned is that following selectivity will work against the latter, since they are very likely to lack the right policies to begin with. The Collier-Dollar selectivity model (2002), which is based on the Burnside-Dollar result, allocates aid to poor countries with good policies, where poverty reduction is the main criterion. Poverty-efficient aid allocations are such that they maximize the number of people that are lifted out of poverty, and poverty-efficient aid will be higher when poverty is higher, per capita income is lower, and the policy environment is better (Collier and Dollar, 2002). Selectivity, which is based on the idea that aid effectiveness depends on the policy environment (following Burnside and Dollar 2000), ignores other factors such as history of past conflicts, the level of democracy, or regime type. And yet, we have seen in the previous section that many factors potentially impact on fragility, and that once taken into account, fragile states might end up with more aid than the selectivity model would provide them.

In order to delve further into the issue of aid effectiveness for fragile countries, we estimate an aid-growth model based on the Burnside-Dollar formulation, namely a standard growth model, augmented with aid and fragility variables as follows:

$$\text{growth}_i = \beta_0 + \beta_1 \text{aid}_i + \beta_2 \text{aid}_i^2 + \beta_3 \text{aid}^* \text{fragility}_i + \beta_4 \varphi_i + \varepsilon_i \quad (2)$$

where Growth is the growth rate of GDP per capita, Aid refers to foreign aid as a percentage of gross national income, Aid^2 reflects diminishing returns to aid because of absorptive capacity constraints, the interactive term Aid*Fragility reflects the idea that aid is less effective in fragile states, and φ is a vector of control variables for growth (initial income,

[19]Aid allocation equations typically control for self-interest, recipient needs, and recipient merits. Here, our concern is determining which of the three main components of state fragility are more likely to influence aid allocation, recognizing that each category contains a number of underlying factors.

TABLE 6 Aid effectiveness

Explanatory variables	All	5>	6>
Constant	5.90 (1.11)	3.80 (0.63)	−0.88 (−0.10)
Aid	0.54** (2.14)	0.57** (2.17)	1.01** (2.99)
Aid*Aid	0.01 (0.31)	−0.01 (−0.01)	−0.01 (−0.63)
Aid*Fragility	−0.08* (−1.93)	−0.07* (−1.79)	−0.12** (−2.49)
Initial Income	−0.43 (−1.11)	−0.32 (−0.70)	−0.47 (−0.56)
Population	0.05 (0.24)	0.12 (0.57)	0.44* (1.69)
Literacy rate	−0.01 (−0.06)	−0.01 (−0.15)	−0.02 (−0.96)
Africa	−0.71 (−1.02)	−0.79 (−1.10)	−0.35 (−0.45)
N	99	84	44
Adj. R-squared	0.13	0.13	0.19

Except where indicated otherwise, the figures in parentheses are the t-values. Standard errors are White-robust.

$^{*}p < 0.1$.

$^{**}p < 0.05$.

initial population and a dummy variable for Africa).[20] As in Table 5, we examine this issue for different samples of fragile states and results are reported in Table 6.

The results are interesting in two ways. First, aid ties to capacity regardless of whether one considers all countries or the most fragile ones; in fact, the coefficients on the aid variable indicate that aid has a larger impact on growth in more fragile countries, ceteris paribus. Second, the interaction term is negative and more significant when the most fragile states are considered.[21]

Conclusions and Policy Recommendations

There is a relative scarcity of empirical work on the relationship between the determinants of state fragility, and of aid allocation. This article has attempted to throw some light on both questions, using the fragility index developed by CIFP, articulated around the ideas of authority, legitimacy, and capacity. The cross-sectional analysis of the determinants of state fragility based on data for the period 1999–2005 showed that fragility is the result of a number of factors, including political and social, that go beyond conflict and economic dimensions, and of which the level of development seemed to be more important. The use of panel data (which would provide more data points and additional degrees of freedom, as well as enable us to control for country and time effects) and the search for better instruments that account for endogeneity represent in our view important areas for future research.

Our examination of aid allocation revealed that authority and capacity are significant factors, whereas regardless of whether a state is fragile or not, it will receive aid independent of whether its legitimacy is weak or strong. Can we conclude, as we argued at the outset,

[20]Compared to other extensive studies of aid effectiveness, which tend to rely on panel data, this is a preliminary step analyzing the issue of aid effectiveness using cross-sectional data, given that we have data on fragility for the past five years until 2005.

[21]These findings clearly deserve further examination, which we for now leave for future research. We have in mind a model using panel data, with additional controls and a policy index a la Burnside-Dollar that will hopefully clarify the aid effectiveness debate for fragile states.

that states become fragile because questions of legitimacy may not be factored into aid programming? To some extent, yes. On the one hand, our examination of the aid-growth relationship showed that aid "works" independent of whether countries are fragile or not. On the other hand, even if we were to assume that aid has a positive impact, one still needs to think about the types of intervention that can take place in fragile environments, beyond just increasing aid flows or increasing authority and capacity structures.

We believe that the ALC framework can serve as a useful point of departure for further research into this question (for example, program vs. project lending, targeting poverty as opposed to governance, or identifying absorptive capacities). Examining fragility using the ALC lens, or different indicator clusters, provides the analyst with a better idea of what areas should be emphasized. Lack of capacity on the part of the state, which is confirmed by the empirical investigation above seems to be important. However, to the extent that this may be correlated with the other components, namely authority and legitimacy, and given that the data show that some countries are more deficient in those sectors, aid focusing more directly on governance or corruption may be more helpful, for example, than direct attacks on poverty. It is also important to note that the general results (or shortcomings) from the literature and experience of the last 50 years as to why aid has not yielded expected results such as the lack of enforcement of conditionality, the failure of aid to "buy" reforms, the volatility of aid flows, lack of absorptive capacity, and fungibility, can all be examined using the ALC framework as a frame of reference. These are areas that need to be investigated further using the approach adopted by recent aid effectiveness studies. Another interesting extension would be to break down fragility into its ALC components and examine the implications of each for aid effectiveness. In sum, we find the ALC approach to be a useful method through which to bring to bear the voluminous, disparate, and in some cases contradictory literature related to state fragility on questions of immediate relevance to policymakers.

References

Alao, C. 1999. The problem of the failed state in Africa. In *International security management and the United Nations,* eds. M. Alagappa and T. Inoguchi. Tokyo: United Nations University Press, pp. 83–102.

Alesina, A., S. Özler, A. Roubini, and P. Swagel. 1996. Political instability and economic growth. *Journal of Economic Growth* 1(2): 189–211.

Baker, P., and J. Ausink.1996. State collapse and ethnic violence: Toward a predictive model. *Parameters* (Spring): 19–31.

Barbieri, K., and G. Schneider. 1999. Globalization and peace: Assessing new directions in the study of trade and conflict. *Journal of Peace Research* 36: 387–404.

Bauer, P. 1981. *Equality, the Third World and economic delusion*. Weidenfeld: London.

Berger, S. 2000. A foreign policy for the global age. *Foreign Affairs* (November/December).

Boone, P. 1996. Politics and the effectiveness of foreign aid. *European Economic Review* 40: 289–329.

Boutros-Ghali, B. 1992. *An agenda for peace: Preventive diplomacy, peacemaking and peace-keeping* (Report of the Secretary-General pursuant to the statement adopted by the Summit Meeting of the Security Council on 31 January 1992), New York: United Nations, Department of Public Information.

Brecher, M., and J. Wilkenfeld. 1989. *Crisis, conflict and instability*. Oxford: Pergamon Press.

Burnside, C., and D. Dollar. 1997. Aid, policies and growth, Policy Research Working Paper 1777, The World Bank.

Burnside, C., and D. Dollar. 2000. Aid, policies and growth. *American Economic Review* 90(4): 847–868 (September).

Carment, D., S. El-Achkar, S. Prest, and Y. Samy. 2006. The 2006 country indicators for foreign policy: Opportunities and challenges for Canada. *Canadian Foreign Policy* 13(1): 1–35.

Carnegie Commission on Preventing Deadly Conflict. 1997. *Final report*. Washington DC: Carnegie Corporation of New York.

Choucri, N., E. Christi, D. Goldsmith, D. Mistree, S. E. Madnick, J. B. Morrison, M. D. Siegel, and M. Sweitzer-Hamilton. 2006. Understanding & modeling state stability: Exploiting system dynamics. *MIT Sloan Working Paper* 4574–06, *CISL Working Paper* 2006–02, January.

Chauvet, L., and P. Collier. 2004. *Development effectiveness in fragile states: Spillovers and turnarounds*. Oxford: Centre for the Study of African Economies, Department of Economics, Oxford University (January). Available online: www.oecd.org/dataoecd/32/59/34255628.pdf (Accessed 1 September 2006).

Clemens, M., S. Radelet, and R. Bhavnani. 2004. Counting chickens when they hatch: The short-term effect of aid on growth. *Centre for Global Development Working Paper* 44: 1–76.

Collier, P., V. L. Elliot, H. Hegre, A. Hoeffler, M. Reynal-Querol, and N. Sambanis. 2003. *Breaking the conflict trap: Civil war and development policy,* vol. 1. Washington, DC: The World Bank.

Collier, P., and L. Chauvet. 2005. *Development effectiveness in fragile states: Spillovers and turnarounds*. Centre for the Study of African Economics, Oxford University, January.

Collier, P., and D. Dollar. 2002. Aid allocation and poverty reduction. *European Economic Review* 26: 1475–1500.

Collier, P., and A. Hoeffler. 1998. On economic causes of civil war. *Oxford Economic Papers* 50: 563–573.

Collier, P. 2007. *The bottom billion: Why the poorest countries are failing and what can be done about it*. New York: Oxford University Press.

Cramer, C. 2005. Inequality and conflict: A review of an age-old concern. Identities, Conflict and Cohesion Programme, United Nations Research Institute for Social Development, Paper Number 11.

Dahl, R. A. 1971. *Polyarchy: Participation and opposition*. New Haven: Yale University Press.

Diamond, J. 1997. *Guns, germs, and steel: The fates of human societies*. New York: W. W. Norton.

Dunning, T. 2005. Resource dependence, economic performance, and political stability. *Journal of Conflict Resolution* 49: 451–482.

Easterly, W. 2003. Can foreign aid buy growth? *Journal of Economic Perspectives* 17(3): 23–48.

Eckstein, H. 1966. *Divisions and cohesion in democracy: A study of Norway*. Princeton: Princeton University Press.

Esty, D., J. Goldstone, T. R. Gurr, P. Surko, and A. Unger.1995. *Working papers: State failure task force report*. McLean, VA: Science Applications International Corporation.

Esty, D., J. Goldstone, T. R. Gurr, B. Harff, M. Levy, G. D. Dabelko, P. Surko, and A. N. Unger. 1998. *State failure task force report: Phase II findings*. McLean, VA: Science Applications International Corporation.

Fearon, J., and D. Laitin. 2003. Ethnicity, insurgency and civil war. *American Political Science Review* 97 (February): 75–90.

Fein, H. 1995. *Life-integrity violation analysis and democracy in 145 countries and regions, 1980–1987*. Ann Arbor: University of Michigan, the Inter-university Consortium for Political and Social Research.

Feng, Y. 1996. Democracy, political stability and economic growth. *British Journal of Political Science* 27: 391–418.

Fund for Peace. 2005. *The failed states index*. Available online: www.fundforpeace.org/, accessed 15 Nov 2007.

Fund for Peace. 2006. The failed states index. *Foreign Policy* (May/Jun): 50–54.

Fund for Peace. 2007. The failed states index 2007. *Foreign Policy* (July/Aug): 54–63.

Goldstone, J., R. Bates, T. R. Gurr, M. Lustik, M. G. Marshall, J. Ulfelder, and M. Woodward. 2005. A global forecasting model of political instability. Paper prepared for presentation at the Annual Meeting of the American Political Science Association, Washington, DC, September 1–4.

Gore, C. 2000. The rise and fall of the Washington consensus as a paradigm for developing countries. *World Development* 28(5): 789–804.

Gupta, D. 1990. *The economics of political violence: The effect of political instability on economic growth*. New York: Praeger.

Gupta, K. L., and M. A. Islam,1983. *Foreign capital, savings and growth—an international cross-section study*. Dordrecht: Reidel Publishing Company.

Gurr, T. R. 1993. Why minorities rebel: A global analysis of communal mobilization and conflict since 1945. *International Political Science Review* (April) 14: 161–201.

Gurr, T. R. 1970. *Why men rebel*. Princeton: Princeton University Press.

Hansen, H., and F. Tarp 2000. Aid effectiveness disputed. *Journal of International Development* 12: 375–398.

Hegre, H., T. Ellingsen, S. Gates, and N. P. Gleditsch. 2001. Toward a democratic civil peace? Democracy, political change, and civil war, 1816–1992. *American Political Science Review* 95 (March): 33–48.

Herbst, J. 1996/7. Responding to state failure in Africa. *International Security* 21: 120–144.

Hewitt, J., J. Wilkenfeld, and T. R. Gurr. 2007. *Peace and conflict 2008*. Boulder, CO: Paradigm Publishers.

Holsti, K. 1995. War, peace, and the state of the state. *International Political Science Review* 16: 319–339.

Human Security Centre. 2005. *Human security report*. University of British Columbia, Available online: http://www.humansecurityreport.info/, accessed 18 November 2005.

Hurwitz, L. 1973. Contemporary approaches to political stability. *Comparative Politics* 5(3): 449–463.

International Commission on Intervention and State Sovereignty (ICISS). 2001. *The responsibility to protect*. Ottawa: International Development Research Centre.

Kant, I. [1795] 2003. *To perpetual peace: A philosophical sketch*. Trans. Ted Humphrey. Indianapolis: Hackett Publishing.

Kingwell, M. 2007. What fragility can teach us. *The Globe and Mail*, F7, October 27.

Lake, D., and D. Rothchild, eds. 1998. *The international spread of ethnic conflict: Fear, diffusion, and escalation*. Princeton: Princeton University Press.

Lijphart, A. 1968. *The politics of accommodation: Pluralism and democracy in the Netherlands*. Berkeley and Los Angeles: University of California Press.

Luckham, R., I. Ahmed, R. Muggah, and S. White. 2001. Conflict and poverty in sub-Saharan Africa: An assessment of the issues and evidence. Institute of Development Studies, Working Paper 128, University of Sussex.

Mansfield, E., and J. Snyder. 1995. Democratization and war. *Foreign Affairs* (May/June) 74: 79–97.

Marshall, M., and J. Goldstone. 2007. Global report on conflict, governance, and state fragility 2007: Gauging system performance and fragility in the globalization era. *Foreign Policy Bulletin* 17(1): 3–21.

McGillivray, M. 2007. State fragility and aid effectiveness: Classification implications. Paper Presented at the UNU–Wider Conference on Fragile States–Fragile Groups, 15–16 June, Helsinki.

Morcos, K. 2005. Chair's summary. Remarks presented at the Senior Level Forum on Development Effectiveness in Fragile States, London, 13–14 January. Available online: http://www.oecd.org/dataoecd/60/37/34401185.pdf (Accessed 1 September 2006).

Mosley, P. 1980. Aid, savings and growth revisited. *Oxford Bulletin of Economics and Statistics* 42(2): 79–95.

Mosley, P. 1986. Aid effectiveness: The micro-macro paradox. *Institute of Development Studies Bulletin* 17: 214–225.

Moreno-Torres, M., and M. Anderson 2004. Fragile states: Defining difficult environments for poverty reduction. United Kingdom Department for International Cooperation, PRDE Working Paper 1 (August). Available online: http://siteresources.worldbank.org/INTLICUS/64137341-

1094571451760/20357055/PRDE_WP_1%20Defining%20fragile%20states.pdf (accessed 22 September 2008).
Muller, E. 1985. Income inequality, regime repressiveness and political violence. *American Sociological Review* 50(1): 47–61.
Nairn, T. 1977. *The break-up of Britain: Crisis and neo-nationalism*. Glasgow: Big Thinking.
OECD. 2001. Poor performers: Basic approaches for supporting development in difficult partnerships—Note by the Secretariat. Organization for Economic Cooperation and Development. OECD DAC (November). Available online: www.oecd.org/dataoecd/26/56/21684456.pdf, accessed 1 September 2006.
OECD. 2005. Paris declaration on aid effectiveness: Ownership, harmonisation, alignment, results and mutual accountability. *High Level Forum*, Paris, 28 February–2 March, 2005, available online: www.oecd.org/dataoecd/11/41/34428351.pdf (accessed 22, September 2008).
Papanek, G. 1973. Aid, foreign private investment, savings, and growth in less developed countries. *The Journal of Political Economy* 81(1): 120–130.
Ramey, G., and V. Ramey. 1995. Cross-country evidence on the link between volatility and growth. *American Economic Review* 85(5): 1138–1151.
Reuveny, R. 2000. The trade and conflict debate: A survey of theory, evidence and future research. *Peace Economics, Peace Science and Public Policy* 6(1): 1–49.
Rotberg, R. 2003. The failure and collapse of nation-states: Breakdown, prevention, and repair. In *When states fail: Causes and consequences*, ed. R. Rotberg. Princeton: Princeton University Press.
Russett, B. and, J. Oneal. 1999. Assessing the liberal peace with alternative specifications: Trade still reduces conflict. *Journal of Peace Research* 36(4): 423–442.
Stedman, S. 1996. Conflict and conciliation in sub-Saharan Africa. In *The international dimensions of internal conflict*, ed. M. Brown. Cambridge: MIT Press.
Stiglitz, J. 1998. More instruments and broader goals: Moving toward the post–Washington consensus. The 1998 WIDER Annual Lecture, Helsinki, Finland, January 7.
Tilly, C. 1978. *From mobilization to revolution*. Reading: Addison-Wesley.
Tilly, C. 2003. *The politics of collective violence*. Cambridge: Cambridge University Press.
United Kingdom. 2005. Investing in prevention: An international strategy to manage risks of instability and improve crisis response: A prime minister's strategy unit report to the government of the UK (February). Available online: www.strategy.gov.uk/work_areas/countries_at_risk/index.asp, accessed 1 September 2006.
United Nations. 2004. A more secure world: Our shared responsibility. Report of the Secretary General's high-level panel on threats, challenges and change. Available online: http://www.un.org/secureworld/, accessed 18 November 2005.
United States. 2005. *Fragile states strategy*. USAID. Available online: www.usaid.gov/policy/2005_fragile_states_strategy.pdf. USAID, accessed 15 November 2007.
Uphoff, N. 1989. Distinguishing power, authority & legitimacy: Taking Max Weber at his word by using resources-exchange analysis. *Polity* 22(2): 295–322.
Urdal, H. 2004. The devil is in the demographics: The effect of youth bulges on domestic armed conflict 1950–2000. Washington DC: World Bank Social Development Papers, Paper 14.
Van de Walle, N. 2001. The economic correlates of state collapse. Paper presented at a Conference on State Failure, Harvard University, 10 January.
Weber, M. [1922] 1968. *Economy and society*, eds. Guenter Roth and Claus Wittich. New York: Bedminster Press, republished by University of California Press, Berkeley CA, 1978.
World Bank. 2002. *World Bank Group work in low-income countries under stress: A task force report*. Low Income Countries Under Stress Task Force (September). Available online: http://siteresources.worldbank.org/INTLICUS/Resources/388758-1094226297907/Task_Force_Report.pdf, accessed 20 September 2008.

Appendix

TABLE 1A CIFP's top 40 fragile states

		ALC Scores			Cross-Cutting	Indicator Clusters					
Country	Fragility Index	Authority	Legitimacy	Capacity	Theme Gender	Governance	Economics	Security	Human Development	Demography	Environment
Burundi	8.25	8.04	7.58	8.65	7.42	7.18	8.08	9.17	8.89	7.25	8.00
Congo (Kinshasa)	8.11	7.93	7.58	8.49	7.72	7.67	6.93	9.15	9.70	7.35	5.47
Afghanistan	7.89	9.06	8.42	6.68	–	9.56	6.00	9.53	7.78	7.57	4.33
Somalia	7.86	7.53	8.41	7.82	–	8.90	8.42	7.18	8.51	7.34	7.13
Liberia	7.84	6.18	8.82	8.64	8.52	9.22	7.58	7.20	8.91	6.78	5.40
Chad	7.81	6.79	8.13	8.43	9.33	7.96	7.04	6.89	9.83	7.57	4.87
Ethiopia	7.81	7.58	7.14	8.31	7.47	6.59	7.44	8.07	8.83	8.35	6.40
Cote d'Ivoire	7.79	7.74	7.89	7.79	8.51	7.83	7.09	7.46	8.64	8.15	6.40
Angola	7.73	7.98	7.66	7.55	6.62	7.62	7.21	7.88	9.28	7.58	4.00
Eritrea	7.73	7.04	7.91	8.14	7.00	6.93	7.45	7.68	9.02	7.49	6.07
Haiti	7.72	6.81	8.53	7.94	7.27	8.32	7.24	8.05	7.95	6.90	7.67
Kenya	7.60	7.46	7.68	7.66	8.60	7.32	7.25	6.98	8.40	8.30	6.67
Rwanda	7.55	6.27	7.47	8.51	6.42	6.93	6.74	6.47	8.69	8.43	8.20
Zimbabwe	7.54	6.77	8.33	7.76	7.62	7.49	8.21	6.79	8.40	6.05	6.27
Guinea-Bissau	7.52	6.66	7.42	8.25	8.38	6.93	8.11	5.43	8.60	8.40	4.67
Sierra Leone	7.50	6.55	7.22	8.46	7.60	7.38	8.18	5.70	8.46	7.33	6.00
Congo (Braz.)	7.49	6.70	7.57	8.02	7.06	7.68	7.47	6.69	8.17	8.23	4.20
Sudan	7.48	7.83	7.58	7.21	7.82	7.13	6.38	9.22	8.22	6.95	6.00
West Bank/Gaza	7.41	6.69	10.33	7.50	8.30	6.85	9.08	8.16	4.78	7.00	9.00
Nepal	7.37	6.58	7.76	7.71	7.42	7.63	6.69	8.28	7.73	7.34	6.00
Nigeria	7.33	7.19	7.46	7.37	7.64	7.19	6.65	7.02	8.08	8.30	6.67
Niger	7.28	5.61	7.09	8.63	9.07	6.92	7.58	3.22	9.16	7.70	6.67
Yemen	7.27	6.59	8.32	7.31	8.93	8.00	6.56	7.44	7.20	7.63	8.33

(Continued on next page)

TABLE 1A CIFP's top 40 fragile states *(continued)*

		ALC Scores			Cross-Cutting	Indicator Clusters					
Country	Fragility Index	Authority	Legitimacy	Capacity	Theme Gender	Governance	Economics	Security	Human Development	Demography	Environment
Uganda	7.24	7.38	6.50	7.51	5.33	6.51	6.11	7.38	8.27	8.95	6.67
Central Af. Rep.	7.17	5.47	8.19	7.97	8.33	7.91	7.49	4.96	8.58	7.23	2.67
Mauritania	7.16	5.99	7.81	7.69	9.34	7.64	6.89	5.67	8.23	6.68	6.93
Guinea	7.15	5.97	7.56	7.92	7.40	7.40	7.36	4.87	8.94	6.90	4.93
Burkina Faso	7.00	5.50	6.39	8.28	7.90	5.61	7.16	3.25	8.94	8.40	5.00
Iraq	6.94	7.52	7.50	6.15	6.42	7.60	7.80	9.38	5.53	6.30	4.33
Tanzania	6.90	6.48	6.14	7.61	6.74	5.99	6.23	5.85	9.16	7.28	5.33
Malawi	6.89	5.87	6.29	7.90	7.42	5.78	7.84	3.52	8.45	8.43	7.00
Togo	6.83	5.50	7.54	7.48	8.17	7.56	6.56	4.26	8.21	6.98	6.33
Pakistan	6.82	7.08	6.43	6.83	6.11	6.47	6.01	8.58	6.44	7.34	6.73
Madagascar	6.81	5.06	7.24	7.83	7.94	6.68	6.88	4.89	8.15	7.74	5.00
Mozambique	6.79	5.53	6.12	8.05	6.80	5.05	7.23	5.05	9.20	7.15	3.67
Myanmar	6.79	6.96	7.15	6.47	6.25	6.75	6.96	8.81	6.75	5.20	4.73
Bangladesh	6.77	6.25	7.68	6.72	7.76	8.25	5.77	7.68	6.48	7.03	4.33
Cameroon	6.77	6.02	7.06	7.23	6.60	7.28	6.52	5.56	7.81	7.48	4.33
Mali	6.76	5.40	6.34	8.01	9.02	5.73	7.13	4.14	9.34	7.20	4.33
Laos	6.67	5.83	7.05	7.15	6.16	5.93	7.11	6.41	7.14	7.00	3.67

TABLE 2A Fragility index scoring scale

Score	Description
1–3.5	Country performing well relative to others
3.5–6.5	Country performing at or around the median
6.5+	Country performing poorly relative to others
Highest 5%	Country among worst global performers

TABLE 3A Correlation matrix

	ETRISK	FRG	GDPPC	GINI	HDI	HREM	INMO	LDEM	LIT	POPDIVET	POPDIVREL	TRAD	UNEM
ETRISK	1.00												
FRG	0.09	1.00											
GDPPC	−0.02	−0.71	1.00										
GINI	−0.19	0.12	0.04	1.00									
HDI	−0.07	−0.85	0.72	−0.02	1.00								
HREM	−0.06	−0.48	0.42	0.42	0.54	1.00							
INMO	0.08	0.79	−0.58	0.06	−0.93	−0.51	1.00						
LDEM	0.18	−0.50	0.45	0.22	0.56	0.75	−0.56	1.00					
LIT	−0.16	−0.71	0.51	0.04	0.84	0.51	−0.76	0.55	1.00				
POPDIVET	0.27	0.48	−0.23	−0.04	−0.36	−0.33	0.40	−0.15	−0.14	1.00			
POPDIVREL	0.03	0.21	−0.08	−0.19	−0.30	−0.09	0.32	−0.27	−0.12	0.41	1.00		
TRAD	−0.25	−0.35	0.06	−0.21	0.16	0.11	−0.22	0.12	0.33	−0.23	−0.02	1.00	
UNEM	−0.16	0.05	0.16	0.17	−0.11	0.08	0.06	−0.09	−0.16	−0.23	0.07	0.03	1.00

Temporal Analysis of Political Instability through Descriptive Subgroup Discovery

DANIEL LAMBACH
DRAGAN GAMBERGER

Introduction

The study of intrastate conflict, from revolutions and coups to ethnic conflict, civil wars, and genocide, has been one of the most active fields in the post-World War II, and especially the post–Cold War social sciences. As a result, our knowledge about the onset and duration of such crises has increased markedly. Recently, the field has come to be dominated by quantitative studies, which have yielded impressive results about factors correlating closely with conflict. These studies, whether they employ regression models, econometric approaches, or other statistical methods, assume that the cases they code as "civil wars" (or whatever the dependent variable) are sufficiently alike to merit being grouped in a single category, and that all cases with a given outcome have developed along the same lines. This raises two potential methodological problems. First, can we be sure that all cases that are coded the same on their outcome variable are actually alike, and if not, how do we determine clusters of distinctive cases? And second, even if the cases are similar, can we rule out the possibility of equifinality, i.e., that different cases might have developed along different paths but ended up in the same state (George & Bennett, 2005)?

This paper presents results of an effort to apply the methodology of knowledge discovery in databases for the qualitative analysis of social science data. Specifically, we use the recently developed method of subgroup discovery to solve the two methodological problems mentioned above. We employ this methodology to analyze data previously gathered by the Political Instability Task Force (PITF) to identify potential correlates of political instability. In the next section, we will give an overview of current literature on the causes and correlates of intrastate conflict, including the findings of the PITF. We then describe the data set and introduce the methodology. Subsequently, we discuss several possible descriptive models for the onset of political instability that emerge from the analysis. Finally, we offer some tentative conclusions drawn from these findings.

Review of the Literature

Current quantitative literature on intrastate conflict (as an umbrella term for ethnic conflicts, civil war, and similar phenomena of domestic political instability) is primarily concerned with the explanation of the onset, the severity (Lacina, 2006) and the duration of conflict (Fearon, 2004). A number of factors have been identified as positively affecting conflict onset. These include a lack of state capacity (Fearon & Laitin, 2003), low economic prosperity, lootable resources (Collier & Hoeffler, 2001), population size (Sambanis, 2004), earlier conflict in the same country or ongoing conflicts in neighboring countries (Esty et al., 1998b), and a "hybrid" regime type (Hegre et al., 2001). Population density and distance to the capital city can also affect conflict risk. However, the influence of these last variables is conditioned by the type of conflict (Buhaug & Rød, 2006). In contrast, several variables are considered either to be insignificant or to produce mixed results, among them ethno-linguistic fractionalization and rough terrain (Fearon & Laitin, 2003; Buhaug & Rød, 2006).

As Sambanis (2004) has pointed out, by employing different data sets using varying definitions of their dependent variable, these studies have sometimes come to starkly different conclusions. Instances of civil war that are coded in one data set are missing in another. Sambanis suggests that this might have skewed results, for example with regard to ethno-linguistic fractionalization: "The results presented here show that the estimated coefficients of most variables vary widely as a result of changes in the coded onset of civil war" (Sambanis, 2004: 855).

These studies, just like the PITF, employ static variables as causal factors for the explanation of conflict onset. Changes in these variables are implicitly recorded in new values for the next country-year cases in the data set. With very few exceptions (GDP growth), dynamic indicators per se are not considered to be relevant independent variables. This is in stark contrast to earlier literature on the social and political roots of conflict. Huntington, for example, theorized that political conflict may result from the modernization of society, which creates new demands for participation exceeding the institutional capacity of the political system (Huntington, 1969: 4).

The Political Instability Task Force

The PITF was founded in 1994 under the name State Failure Task Force (SFTF) at the request of the United States government (Esty et al., 1998a). The SFTF's original aim, as the founding name implied, had been to identify the root causes of state failure. However, preliminary research showed that only 18 episodes of state failure had occurred in the initial time frame of 1955–1994, too few for meaningful quantitative analysis (Esty et al., 1995). As a result, the Task Force broadened the concept of state failure so that it included a wider range of phenomena, namely, adverse regime changes, genocides/politicides, revolutionary wars, and ethnic conflict. The SFTF acknowledged that this differed from the prior definition of state failure by renaming itself the Political Instability Task Force in 2003.

In its early stages, the PITF[1] compiled a data set encompassing 195 countries in the period from 1955 to 1994 (later updated to include events up to 2004).[2] Each country was described by 87 political, demographic, economic, social, and environmental variables that had been collected from different sources, including the World Bank and United Nations. In the above period, 131 instances of "state failure" (hereafter referred to as "positive cases") had been recorded in 96 different countries. Since the PITF was trying to construct a predictive model, data from 2 years prior to the instability was combined with the outcome variable. The same procedure was employed for constructing negative (control) cases, three of which were added to the data set for each positive case. Control cases were defined as not having experienced instability during the 5-year period immediately preceding and following the sampling year.

The goal of the PITF is the development of statistical models that can be used to identify countries at greater risk of political instability (Esty et al., 1998b). Initially, single-variable correlations were used to detect potentially significant variables that were then subjected to multivariate logistic regression and neural network modeling. In its Phase I report, the Task Force identified a combination of three variables as providing the best predictive power: high infant mortality, partial democracy, and low trade openness. This global model was able to predict the positive cases with an accuracy of 60% (Esty et al., 1995). In the second and third phases, this model was refined by adding further independent variables, namely a large population and high population density as well as two or more bordering states with major civil conflict. Including these variables improved the model's accuracy to 72%. There are differences in the predictive power of these six variables, with quality of life (for which infant mortality is a proxy) and regime type being the most robust. In addition, the PITF developed special models for specific outcome types (e.g., ethnic war) or selected geographical areas (e.g., sub-Saharan Africa, Muslim countries). These models usually attained a level of accuracy around 80%. However, with regard to the global model, it is doubtful whether substantial improvements in prediction accuracy can be realized with this data set and methodology without further loss of parsimony.

Despite these achievements, the PITF has been criticized on theoretical and methodological grounds. On a theoretical level, the most obvious charge against the PITF has long been that what it was studying was not state failure at all. This concept of state failure describes a breakdown of political order and state institutions and is at once both wider and quite different from the four kinds of political instability that the PITF included (Zartman, 1995). For example, state failure theorists would in most cases reject the notion that adverse regime change be considered as a positive case (unless it was particularly violent and took a long time to take effect). The PITF admitted as much in its first report (Esty et al., 1995), but continued to denote its positive cases "state failures," thereby contributing to the confusion. The issue was further obscured by the PITF's unwillingness to provide a clear definition of its dependent variable beyond a simple typology of events classified as state failure.

Even if this critique had merely been an issue of semantics, the lack of a theoretical core led to further problems. King and Zeng (2001) have pointed out that the different kinds of positive cases (adverse regime change, revolutionary war, ethnic war, genocide) actually were discrete types of events with quite distinct causes. This critique was somewhat countered by the PITF's development of separate causal models for these outcome types.

[1]To facilitate reading, we will refer to the task force as PITF even when talking about its work before its re-christening in 2003.

[2]In an intermediate step, the data set had been updated until 1999. This was the most current version of the data set when work on our project began, therefore the data set underlying the present paper covers events between 1955 and 1999. A public version of the data set, including documentation as well as all three reports is available on the PITF website at http://globalpolicy.gmu.edu/pitf/pitfdata.htm.

However, subsuming analytically distinct events into a single category at best reduces theoretical leverage; at worst, it produces misleading results (Sartori, 1970). It raises the danger of equifinality, since differences between positive cases can stem from divergent causal structures. One way to address this problem would be to disaggregate the dependent variable into clusters of related cases and then to analyze these clusters separately, as the PITF did through its development of separate models. However, this strategy then raises the question how such clusters should be determined. The PITF opted for a deductive approach, i.e., it defined the four kinds of positive outcomes on theoretical grounds. Such an approach again carries the risk for concept misformation: how can we be sure that, for example, revolutionary war and ethnic war are distinct groups of cases, or that all cases of adverse regime change are sufficiently alike to be considered the same class of events?

Other lines of criticism have pointed out the limitations of the PITF's approach. Schrodt (2002) criticized the PITF and similar data-mining efforts as only producing correlations without addressing issues of causation. This point is illustrated by infant mortality—it is clear that infant mortality per se does not cause political violence. However, it is a good indicator for conditions of social squalor, which, in turn, are potential sources of conflict. And while causality and explanation were never on the PITF's agenda, it is important to be aware of this shortcoming. Finally, the limits of the global model are highlighted when looking at specific cases. Daun (2003) has conducted an independent single-case study of the representation of Colombia in the PITF data set: 4 country-years were included as either positive or negative outcomes; in three of these cases, the global model predicted the wrong result. This is a general weakness of the global model that is especially pronounced in "marginal" cases like Colombia, where violence takes a less ubiquitous form than, for example, in Somalia.

Subgroup Discovery Methodology

Cases in our analysis are represented by periods of 5 country-years without political instability. Positive cases encompass those periods immediately preceding a year with a recognized instability, while negative cases are those where political instability did not occur for at least another 5 years after the conclusion of the initial 5-year period. From each of the 87 variables available in the data set per country-year, we constructed 11 descriptors representing the 5-year time sequence. They are: mean value over 5 years; minimum value; maximum value; slope; standard deviation; range (maximum − minimum value); positive distortion (maximum − mean value); negative distortion (mean − minimum value); temporal distance between extremes (position of the minimum − position of the maximum value in years); relative distance of the minimum value in respect to the beginning of the time sequence; and relative distance of the maximum value in respect to the beginning of the time sequence.

These descriptors reflect both absolute (static) properties of the input variable (mean, maximum, minimum values, range, distortions), and temporal (dynamic) properties of the changes (slope, positions of the extremes, standard deviation). The result is that, due to the described data preparation phase, each case is represented by 957 numerical descriptors. The subgroup discovery algorithm that follows uses only these descriptor values as input.[3]

The set of cases consists of the group of positive cases P (countries experiencing political instability) and a control group of negative cases N (countries not experiencing

[3]It should be noted that the advantage of the applied subgroup discovery approach is that, due to the built-in anti-overfitting measures, the methodology can accept a large number of input variables even when the number of cases is relatively modest. For example, in the functional genomics domain successful experiments have been done with as much as 16,000 descriptors (Gamberger et al., 2004). In contrast to multivariate logistic regression modeling that needs to have explicit single-variable

political instability). The goal of the subgroup discovery algorithm is to construct rules that are true for positive cases and false for negative cases. It is not necessary that rules are true for all positive cases and false for all negative cases, but the intention is to find short rules that are true for large subsets of positive examples and at the same time that are false for large subsets of negative examples. Subgroup sizes are not defined in advance but the algorithm tends to make them as large as possible. A rule with ideal covering properties is true for all positive cases and not true for all negative ones. Positive cases covered by a rule are also called "true positives" and denoted by TP, while negative cases covered by the rule are called "false positives" (FP). All remaining negative cases not covered by the rule are called "true negatives" (TN). An ideal rule has TP = P and TN = N, and because of N = TN + FP the ideal rule has FP = 0.

The first step in the rule construction process is the construction of all possible "features" representing elementary rule building blocks. For descriptors on an interval scale the features have the form Descriptor > value or Descriptor < value. Examples from the PITF data set are 'mean population density > 40.8 or slope of change of GDP per capita < 0.15. For each input descriptor, there can be many different features and the process of their construction is well defined. Practically for each pair of one positive and one negative case it is possible to construct one feature for every descriptor. For example, if we have a positive case with mean population density = 30 and a negative case with 'mean population density = 35, then a feature 'mean population density < 32.5 may be constructed. This feature will successfully discriminate between these two cases because it is true for the positive case and false for the negative one. Many features constructed this way can be immediately discarded if they are only true for a small number of positive cases or if they are false for a small number of negative cases. The reason is that such features are potentially bad rule building blocks because they may lead to results that are overfitted to the available data set and do not describe general relationships. All other features with reasonably good covering properties enter the rule construction process. For the PITF data set, we only accepted features that were true for at least 10 positive cases and, at the same time, that were false for 10 to 15 negative cases, depending on the total number of negative cases.

The central part of the rule construction process is the optimization algorithm searching for combinations of features with optimal covering properties on the given set of cases. It is assumed that features can be connected only by logical conjunction. This means that a combination of features is true for a case only if all features are true for the case and that a combination of features is false for a case if any of the features is false for it. In the subgroup discovery approach, the following rule quality measure Q is used as the optimization goal in the heuristic search of rules:

$$Q = TP/(FP + g)$$

where g is an appropriately selected generalization parameter. High quality rules will have a large Q value and they will cover most positive cases (large TP) and a low number of negative cases (small FP). The number of negative cases tolerated, relative to the number of positive ones covered by the rule, is determined by the parameter g. Rules presented in this work have been induced with g being in the range between 5 and 20. The quality of constructed rules is measured by two values: *sensitivity* which is equal to TP/P and which represents the proportion of positive cases classified as true positives, and *specificity* which

correlation tests in the preprocessing stage in order to reduce the number of inputs to ensure the construction of reasonable models, this approach allows even further expansion of the number of input descriptors.

is equal to TN/N and which represents the proportion of negative cases classified as true negatives. Very good rules have high sensitivity and high specificity, but rules can also be useful if only one of the values is very high.

An example of a real rule constructed for the PITF domain consisting of two features is the following: position of the maximum population value > 2.5 (maximum population is reached in the second half of the 5-year cycle) and 'difference between maximum and minimum values of infant mortality > 7.2. This rule has a sensitivity of 42% (it is true for 50 of 118 positive examples) and a specificity of 78% (it is false for 179 of 229 negative examples where data for both descriptor values is present). This rule is actually the base for the model F1 presented in this work. The transformation from the rule form into the descriptive form is the result of a statistical comparison of the subgroup of 50 positive cases described by the rule on the one hand and all available negative cases on the other. As expected, all positive examples that are true for the rule have increasing population (maximum value near the onset of instability) and large differences in infant mortality in a 5-year period. Moreover, these cases have also significantly decreasing infant mortality characterized by a significant negative slope for infant mortality as well as a significant positive slope for total population. Therefore, the main properties of the model are described as "countries characterized by *significantly increasing population* as a result of *rapidly decreasing infant mortality.*" There can also be other significant differences between the subgroup of positive cases and all negative cases. These differences are listed in the model description as additional but not necessary properties. In the current example, model F1 has "decreasing GDP" (detected as the negative mean value of the slope of GDP for selected positive cases in contrast to the positive mean value that characterizes negative cases) and "growing urban population" as additional characteristics.

The other models presented in this work have been constructed the same way. Although the induced rules should have high predictive accuracy characterized by high sensitivity and specificity values, the real relevance of rules lies in the automatic detection of relevant subgroups of positive cases. Analysis of the statistical properties of these subgroups necessitates the evaluation and interpretation of underlying concepts during the transformation from rule form into model descriptions by an expert in the topic under study. Thus, the advantage of the methodology is that it actively includes human background knowledge and experience into the data analysis process. The disadvantage is that the process may result in different model descriptions depending on the expertise and preferences of the involved social scientist. Initially, the process can be influenced by selecting higher or lower values for the generalization parameter, g, during the rule generation process. Later, the expert can significantly influence the model description by highlighting different additional properties of the model. Also, the approach does not eliminate Schrodt's critique of the PITF's approach (see above) in that it still detects correlation and not causation. Nevertheless, the human evaluation of hypotheses leaves enough freedom for experts to try to detect and incorporate causation into theories that are built based on the induced hypotheses. It also allows us to avoid the pitfalls of concept misformation through the inductive creation of subgroups.

The methodology used in this work represents recent advances in the field of intelligent data analysis. It has already been successfully applied in the fields of medicine (Gamberger et al., 2003), functional genomics (Gamberger et al., 2004), and chemistry (Baker et al., 2004), but it is yet to be generally accepted as a standard approach of data analysis. A goal of this study is to demonstrate its applicability to the social sciences and stimulate broader awareness of its potentials, while methodological details are outside the scope of this paper (Gamberger & Lavrač, 2002).

The computational aspect of the methodology is relatively complex because it makes use of a complete Inductive Learning by Logic Minimization (ILLM) system implemented at the Rudjer Boskovic Institute, Zagreb. Outside researchers interested in the application of the methodology can access the Data Mining Server, publicly available at http://dms.irb.hr, which enables the use of the subgroup discovery tool with user-submitted data. The server presents a very simple and user-friendly interface to the data analysis process, but is limited to 250 cases and 50 descriptors for the sake of public availability. To allow for more complex experiments, please contact the authors of the paper.

The Experimental Setting

In the first phase, we experimented with a data set consisting of 118 positive and 259 negative (control) examples. This data set will be referred to as the F set. The positive cases included all examples for which a preceding 5-year period of political stability existed. Selection of negative examples was relatively random but the intention was to include negative cases, where possible, both for countries which had also generated a positive case, as well as a number of negative examples from countries which did not experience political instability, preferably in the same time periods when positive cases had been generated for other countries. The results obtained were very helpful to illustrate the methodology and its potentials, although later evaluation by a political scientist showed that some positive cases had been included in the data set even though they had not been independent states for the entire preceding 5-year period, raising doubts about the reliability of data. Moreover, expert analysis showed that relying on negative examples drawn from stable and developed countries produced biased results that more often reflected other, more general socio-economic differences between these countries and positive cases rather than differences in political stability as such.

To circumvent the problems detected in the first phase, we then designed two additional experiments. Positive cases remained the same in both experiments: There are 87 cases of political instability that satisfied the conditions that the countries had been independent and stable for at least 5 years before political instability set in and that the 5-year period started after 1955. In the first of these experiments, which will be referred to as the M data set, negative cases were constructed using countries that had never experienced political instability. For each such country (99 in total) two negative cases were constructed with randomly chosen 5-year periods, taking care that these two periods are at least 20 years apart. If it was not possible to guarantee a 20-year interval (because of the relatively recent independence of a country), a minimum distance of 10 years was employed, and if this was still not possible, only a single case was constructed for the country. In total, this gave us 181 negative cases for the M set. This set is intended to highlight immanent properties that differentiate unstable countries from stable ones.

In the second experiment, a different data set (the S set) was constructed with the aim of detecting which variables were most successful in predicting why and when similar countries remained stable or became unstable. To do that, the negative cases in the S set were taken only from countries that had experienced political instability at some other point. For each such country, we tried to find two periods of 5 years which could serve as negative cases: one which was before the onset of instability and another one after its termination. For each negative case, a minimum distance of 10 years to the positive case had to be present. If this was not possible, negative cases with a shorter distance (but not less than 5 years) were accepted. For some countries that experienced a particularly long period of instability (or that had only recently become independent), it was only possible to construct a single negative case, or, for a few countries, no negative case at all. However, it was possible to extract a few additional negative cases from countries which had experienced instability but were dropped from the data set because reliable data had not been present for at least 5 years prior to the onset of instability (see above). This produced a total count of 96 negative cases for data set S.

In the following section we present the results from the analysis of the data sets. It should be pointed again that the method of subgroup discovery does not produce hard and fast results, but helps formulate specific hypotheses through the inductive discovery of subgroups of cases sharing characteristics overlooked otherwise. Therefore, the induced models were later evaluated by asking how well they conform to theory, other research findings and expert knowledge on political instability.

General Political Instability Models (Induced from the F Set)

From the data set constructed in the first phase (F set), it was possible to detect increasing population and a change in GDP per capita as the most relevant driving forces for political instability. The following specific models were constructed:

Model F1

Countries characterized by *significantly increasing population* as a result of *rapidly decreasing infant mortality*. Additional, but not necessary, characteristics are decreasing GDP per capita and growing urban population. Such countries are more likely to suffer an adverse regime change than other countries. An example of such a case would be South Korea in 1961. Model sensitivity is 42% and specificity is 78%. This pattern closely resembles theoretical models from the literature on modernizing and developing countries. Numerous forces of social change coincide: growing demographic pressure coupled with urbanization provide for a tight job market and social dislocation, creating the structural conditions for conflict. An economic recession then becomes the proximate cause of violent conflict. This model agrees with the explanations by Huntington outlined above.

Model F2

Small countries with an *increasing number of conflict in bordering states* in combination with *increasing trade openness*. Additional characteristics are small size, a strongly decreasing GDP per capita and decreasing population. The most frequent outcomes are ethnic wars and complex cases (a combination of several different outcome variables). Bosnia in 1992 is a typical representative of this concept. Model sensitivity is 32% and specificity is 87%. F2 does not lend itself to a parsimonious interpretation like F1 does. It is potentially interesting because it does not agree with existing theories of political instability. The combination of the factors "neighboring conflicts" and "trade openness" is very unusual. While such a combination might seem plausible on the surface (a relatively open society "importing" the seeds of conflict), it is more likely that this model is either a collection of individual outlier cases or that it is held together by omitted background variables. Another possibility is that the key causal variable in F2 is the "increasing number of neighboring countries in conflict" one, which had already been found as contributing to conflict by the PITF and in further research (de Waal, 2000: 1–34). In this case, the other descriptor (trade openness) would be an epiphenomenon, i.e., a factor that correlates with the neighboring conflicts variable without there being a relevant connection between the two.

Model F3

Countries characterized by *increasing population* and *increasing GDP per capita* and *increasing cropland*. Additional characteristics are regime type stability and decreasing trade openness. The most likely outcome is an adverse regime change, with Argentina 1973 being an example. Model sensitivity is 19% and specificity is 93%. There are some similarities with F1: firstly, both models include an increasing population. Secondly, while F3 includes an increase in GDP, the countries in F1 exhibit a decrease. This is not the contradiction it may seem—Clément (2004) has indicated that the economic shift in itself is relevant, but not its direction. These similarities again lead us to the issue of social and economic change. The model conforms to Huntington's explanation, in that it looks at a static regime in a situation of change. The regime is unable or unwilling to adapt to the changing circumstances, which leads to a violent confrontation that effects regime change. The descriptors increasing cropland and decreasing trade openness do not add explanatory value. Beyond infrequent conflicts between pastoralists and nomads in some parts of sub-Saharan Africa, there is no theoretical reason why increasing cropland should contribute to conflict. Decreasing trade openness might suggest international isolation of the country, possibly as a result of a crackdown by the government, but this is highly speculative and merely reinforces the importance of the regime variable. Accuracy characteristics and overlapping between models are illustrated in Figure 1.

The conditions used in all three models are based on complex descriptors related to temporal changes of input descriptors. The result justifies the temporal approach to the

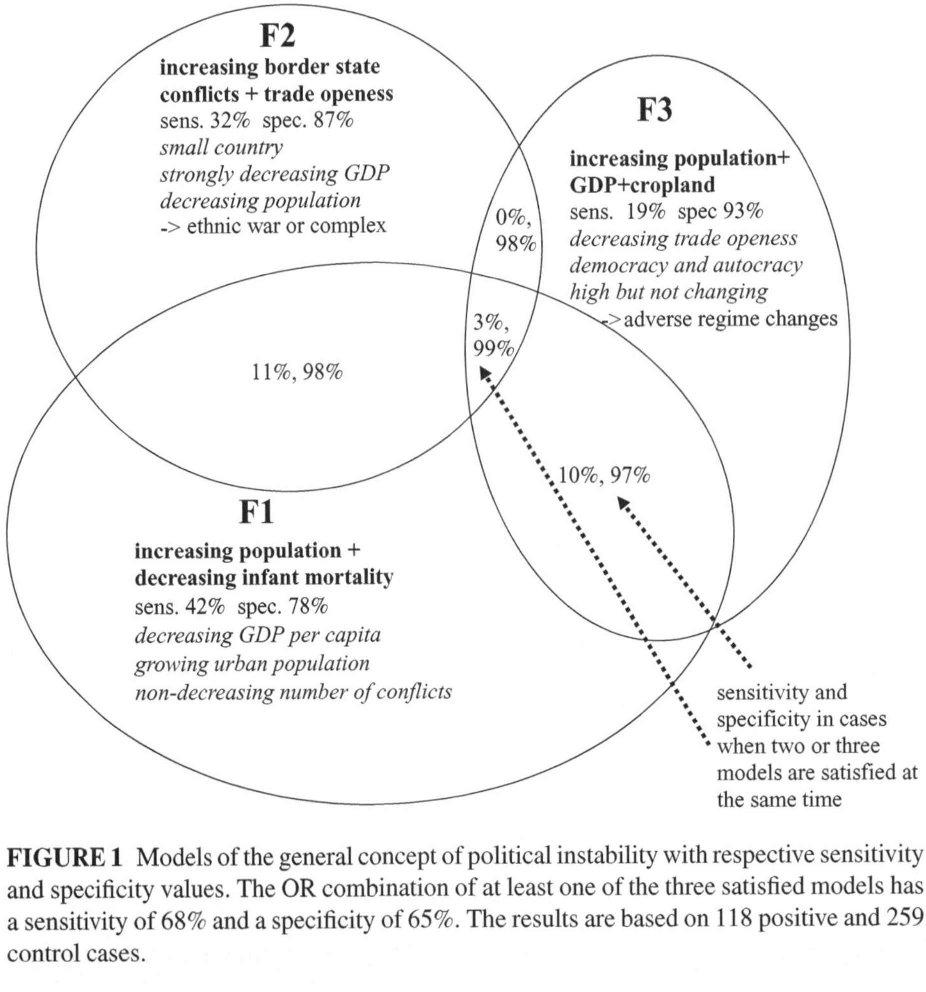

FIGURE 1 Models of the general concept of political instability with respective sensitivity and specificity values. The OR combination of at least one of the three satisfied models has a sensitivity of 68% and a specificity of 65%. The results are based on 118 positive and 259 control cases.

analysis of the PITF data set. But the results allow the conclusion that dynamic conditions are more relevant than the absolute values of these variables in analyzing political instability. For example, "changing quality of life or increasing GDP per capita" can have a much larger impact on political stability than the actual absolute values of infant mortality and GDP per capita. Additionally, some variables have different implications for different models. One example is GDP per capita: in model F3, increasing GDP is a necessary condition while models F1 and F2 both detect decreasing GDP as an additional, but not necessary characteristic of unstable countries. Furthermore, we have an increasing total population in models F1 and F3 as a necessary condition and a decreasing population in model F2 as an additional characteristic. These results demonstrate the importance of partitioning the group of positive cases into subgroups. In other words, different countries seem to follow different paths toward political instability.

Immanent Danger of Political Instability

From the M data set, we induced immanent characteristics of countries that eventually experienced political instability. The most frequently detected properties are an increasing population, a small percentage of urban population, changes in labor force and a not completely democratic government (a Polity score below 7).[4] The following specific models have been detected:

Model M1

Countries characterized by a *very low percentage of urban population (below 41%)* and a *change of leadership*. Additional characteristics are large changes in the total labor force and a changing level of democracy. A typical case of this model is Haiti in1991. Model sensitivity is 25% and specificity is 89%.

Model M2

Countries characterized by *low percentage of urban population (below 55%)* and *changing infant mortality*. Additional characteristics are large changes in total labor force and increasing population. Jordan 1967 represents a good example of this model. Model sensitivity is 62% and specificity is 77%.

Model M3

Countries characterized by *low, but rapidly changing percentage of urban population (below 50%)*. Large changes in the total labor force represent an additional characteristic. One of the cases from M3 is South Africa 1967. Model sensitivity is 49% and specificity is 86%.

Models M1 to M3 conform to Huntington's general proposition that modernizing societies are at a special risk of instability and conflict. Further, low percentage of urban population is a constant property that represents a necessary (if not sufficient) condition. The large segment of rural population indicates that these are underdeveloped and modernizing countries. A political crisis ensues when a proximate cause is added to this structural

[4]The Polity score is drawn from the Polity IV data set. It stands for the difference between the two variables showing the degree of democracy (DEMOC) and autocracy (AUTOC) in a country. Since these have values between 0 and 10, the Polity score ranges from −10 (completely autocratic) to 10 (completely democratic). The Polity IV data set is available at http://www.cidcm.umd.edu/inscr/polity/.

condition, e.g., a change in leadership (M1), a change in the quality of life (M2), or increasing urbanization (M3).

The models M1 to M3, which describe the immanent features of unstable cases, seem very well suited to point out the crucial factors differentiating stable from unstable countries. The models rely more on attributes differentiating developing and developed countries. This is reflected primarily in the attribute "low percentage of urban population" that is present in all models, and the condition "changing labor force," which is a supporting characteristic for these models. In contrast to models induced from the F data set, models describing immanent characteristics of politically unstable states have a somewhat higher accuracy. This shows that socioeconomic descriptors included into the PITF data set are better for differentiating between unstable and stable countries than for building concrete political instability models.

Specific Conditions for the Onset of Instability

In the S set, negative cases were generated only from the countries that had experienced instability in a time period at least 5 years prior to or after the negative case. Hence, we expected that the models induced would reflect specific conditions that contributed to the immediate onset of political instability. The following models have been constructed:

Model S1

Countries characterized by *rapidly changing infant mortality* but *no significant change in GDP per capita*. Additional characteristics are a relatively high maximum level of democracy, and a large agricultural population. This model is exemplified by Gambia 1994. Model sensitivity is 25% and specificity is 94%. The rapid change in infant mortality points toward major changes in the quality of life. A large agricultural population combined with a formal system of democracy and economic stagnation suggests that these are countries that were suffering from an economic crisis, possibly in combination with uneven development. The economic situation might then have been ripe for exploitation by elite actors capitalizing on popular unrest, which eventually led to regime change.

Model S2

Countries characterized by *semi-democracy (Polity score above -5)* and *a change in membership in regional organizations*. Additional characteristics are an increasing GDP and a changing size of forest cover. Albania 1996 is a typical representative of this subgroup. Model sensitivity is 26% and specificity is 93%. Having a democracy score somewhere between 0 and 7 on the Polity scale, these regimes can be considered anocracies. This model most likely captures countries in transition that had previously emerged from single-party or military rule and taken some steps towards a true polyarchy which makes their political system highly vulnerable. The variable "changing membership in regional organizations" is hard to interpret. A changing membership reflects either diplomatic realignment or deepening regional integration. Since semi-democracies are less stable than other regimes types (Hegre et al., 2001), inclusion of this variable means that this model most likely covers a group of countries in transition.

The models induced from the S set have low sensitivity scores and a relatively low level of accuracy. Beyond the regime characteristic (where both models exclude highly autocratic countries), the models have no common features. A possible interpretation for this phenomenon is that generalizing specific conditions for the onset of political instability is a very difficult task. Instead, onset might be influenced more strongly by context- and case-specific factors. To understand the problem of political instability, it is necessary to integrate the models obtained from the S set with conditions detected in set M. This is especially relevant for the conditions based on the type of government. From set M we have learned that nondemocratic countries have a larger risk of instability while the S set points out the relative dangers of nonautocracy. These two specific properties, combined into the general proposition that semi-democracies are at a higher risk of instability, support the findings of the PITF as well as of the quantitative literature on civil war.

Toward a Unified Concept of Political Instability

The induced models and their expert evaluation allow for the extraction of some general conclusions about the causes of political instability. A very simple proposition to be inferred is that instability and its onset are a very complicated process that cannot be predicted solely from static socioeconomic data. In that respect, our results confirm those reported by the PITF. However, the induced models allow for some further theorizing.

Summarizing the results, we can say that instability is characteristic of countries with a high share of rural population undergoing significant social, economic or political change. Fast population growth and a change in GDP are important supporting indicators. A fully developed, "mature" system of democracy seems to able to absorb problems associated with social change, and a high level of democracy is a key characteristic of countries without immanent danger of instability. Onset of political instability usually happens in periods of political transition. This is a fairly trivial observation, but it has to be pointed out that a hybrid regime is not a sufficient condition by itself; it needs to be accompanied by periods of rapid or unbalanced economic and social changes to escalate into crisis. Other proximate risk factors seem to be changes in the country's leadership as well as conflict in neighboring countries.

Our exclusion of positive cases where the country in question had not been independent for at least 5 years before instability set in might have introduced selection bias. However, such bias, if present at all, will likely have caused us to underestimate the robustness of the factors identified. Previous research (Fearon & Laitin, 2003) has shown that countries are at a significantly higher risk for conflict for a few years after independence. While they propose no theory as to why this might be the case, our findings support the notion that this increase is due to the attendant political transition which is accompanied by social and economic change. In addition, newly independent countries since 1945 have usually been overwhelmingly poor which correlates strongly with large population growth. All of these are factors that show up as highly relevant in our model.

Discussion and Conclusions

The relevance of the work presented in this paper lies in the effort to use the novel methodology of subgroup discovery that had not previously been used to analyze social science data. The methodology is able to produce results that go beyond those obtained by other quantitative studies. What the new methodology provides is a) the automatic detection of relevant subconcepts and b) a concise description of subgroups through co-occurring variables.

As regards the first point, the automatic detection of subgroups is relevant whenever a complicated target concept like political instability can be more effectively described by breaking it down into different but potentially nonexclusive subconcepts. The advantage of the subgroup discovery approach is that expert knowledge is not necessary for this step. More importantly, expert-generated subgroups might be inappropriate: in our research, neither regional nor religious properties show up as central features or as descriptive properties of the induced subgroups. As to the type of instability, it should be noted that models F1 and F3 predominantly describe adverse regime changes while F2 mainly corresponds to ethnic wars and complex cases (see Figure 1). This suggests that ethnic wars and adverse regime changes really have different causal backgrounds.

Additionally, different subgroups may have the same relevant variables but with opposite effects. For example, models F1 and F2 identified decreasing GDP per capita as an indicator of crisis while F3 included increasing GDP per capita as a crucial variable. In such cases, the search for globally relevant variables may completely fail to detect such descriptors even though they are relevant for each subconcept. In this way the subgroup discovery approach may prevent such important variables from being overlooked.

The second advantage of this methodology is the explicit construction of models by co-occurring factors in a form directly and easily interpretable by researchers with expertise about cases or the subject matter. In this paper, expert evaluation showed that most models corresponded to existing research results on mass political violence. Even when single variables or whole models were identified as not directly useful these variables or models may be subjected to further analysis, e.g., in the form of process tracing (George & Bennett, 2005). We cannot expect a revolutionary, novel theory about political instability from the methodology at hand, but the results are relevant because they give some added weight to previously existing theories, such as Huntington's theory of political development.

From a methodological point of view, the agreement between induced models and existing research demonstrates that the methodology is able to detect relevant correlations from the available data without expert knowledge. This will be useful for future applications of this methodology, especially for disciplines and problems without a sufficiently developed theoretical background.

The results demonstrate the relevance of the temporal analysis of available data. Although there are variables (like the percentage of urban population or the level of democracy) where absolute or mean values are relevant, a complete understanding of the problem is not possible without analyzing changes in the values of the variables. Changes in total population, GDP and infant mortality seem to have an especially high significance. Interestingly enough, the absolute values of these variables have not been detected as relevant.

The final advantage of the applied methodology is that it can accept a large number of input variables even when the number of cases is modest. In contrast to multivariate logistic regression modeling which needs to have preliminary single-variable correlation tests in order to reduce the number of inputs and ensure the construction of reasonable models, in the subgroup discovery approach a further expansion of the number of input variables is possible. This enables an effective handling of temporal properties of input variables, which may also be relevant for other applications in the social sciences.

Acknowledgments

We thank the journal editor as well as several anonymous reviewers for their very useful comments that significantly helped us in preparing the final version of the paper. This research was supported in part by the NATO *Programme for Security Though Science*.

References

Baker, J. R., D. Gamberger, J. R. Mihelcic, and A. Sabljic. 2004. Evaluation of artificial intelligence based models for chemical biodegradability prediction. *Molecules* 9: 989–1004.

Buhaug, H., and J. K. Rød. 2006. Local determinants of African civil wars, 1970–2001. *Political Geography* 24: 399–418.

Clément, C. 2004. *State collapse: A common causal pattern?* Unpublished Ph.D. dissertation, Université de Louvain, Belgium.

Collier, P., and A. Hoeffler. 2001. *Greed and grievance in civil war.* Washington, DC: World Bank Research Paper, October 2001.

Daun, A. 2003. *Staatszerfall in Kolumbien.* Unpublished M.A. Thesis, University of Cologne, Germany.

de Waal, Alex. 2000. *Who fights? Who cares? War and humanitarian action in Africa.* Trenton, Asmara: Africa World Press.

Esty, D. C., J. A. Goldstone, T. R. Gurr, P. T. Surko, and A. N. Unger. 1995. *State Failure Task Force Phase I Report.* McLean: Science Applications International Corporation.

Esty, D. C., J. A. Goldstone, T. R. Gurr, B. Harff, P. T. Surko, A. N. Unger, and R. Chen. 1998a. The State Failure project: Early warning for US foreign policy planning. In *Preventive measures: Building risk assessment and crisis early warning systems*, ed, 27–38. Lanham, Oxford: Rowman and Littlefield.

Esty, D. C., J. A. Goldstone, T. R. Gurr, B. Harff, M. Levy, G. D. Dabelko, P. T. Surko, and A. N. Unger. 1998b. *State Failure Task Force Phase II Report.* McLean: Science Applications International Corporation.

Fearon, J. D. 2004. Why do some civil wars last so much longer than others? *Journal of Peace Research* 41: 275–301.

Fearon, J. D., and D. D. Laitin. 2003. Ethnicity, insurgency, and civil war. *American Political Science Review* 97: 75–90.

Gamberger, D., and N. Lavrač. 2002. Expert-guided subgroup discovery: Methodology and application. *Journal of Artificial Intelligence Research* 17: 501–527.

Gamberger, D., N. Lavrac, and G. Krstacic. 2003. Active subgroup mining: A case study in a coronary heart disease risk group detection. *Artificial Intelligence in Medicine* 28: 27–57.

Gamberger, D., N. Lavrac, F. Zelezny, and J. Tolar. 2004. Induction of comprehensible models for gene expression data sets by subgroup discovery methodology. *Journal of Biomedical Informatics* 37: 269–284.

George, A., and A. Bennett. 2005. *Case studies and theory development in the social sciences.* Cambridge, London: MIT Press.

Goldstone, J. A. 2000. *State Failure Task Force Phase III Report.* McLean: Science Applications International Corporation.

Hegre, H. et al. 2001. Toward a democratic civil peace? Democracy, political chance, and civil war. *American Political Science Review* 95: 33–48.

Huntington, S. P. 1969. *Political order in changing societies.* New Haven: Yale University Press.

King, G., and L. Zeng. 2001. Improving forecasts for state failure. *World Politics* 53: 623–658.

Lacina, B. 2006. Explaining the severity of civil wars. *Journal of Conflict Resolution* 50: 276–289.

Sambanis, N. 2004. What is civil war? Conceptual and empirical complexities of an operational definition. *Journal of Conflict Resolution* 48: 814–858.

Sartori, G. 1970. Concept misformation in comparative politics. *American Political Science Review* 64: 1033–1053.

Schrodt, P. A. 2002. Forecasts and contingencies: From methodology to policy. Paper prepared for the APSA Conference, Boston, MA, 29 August-1 September 2002.

Zartman, I. W. 1995. Introduction: Posing the problem of state collapse. In I. W. Zartman, ed. *Collapsed states: The disintegration and restoration of legitimate authority.* Boulder: Lynne Rienner.

Failing States and Failing Regimes: The Prediction and Simulation of State Failure

NEIL A. ENGLEHART
MARC V. SIMON

How do we know a state has failed, or is failing? When, for instance, did the Afghan state fail? 1992, 1989, 1978 and 1976 are all plausible dates, depending on one's definition of failure – if indeed it was ever a successful state at all. Did Kenya experience state failure in early 2008? Have Lebanon or Cambodia ceased to be failing states? The answer to these questions obviously depends on what is meant by state failure, and answering them requires operationalizing the concept. Unfortunately, state failure is still inadequately theorized and there is little consensus on how best to operationalize it. The purpose of this chapter is to advance the theory and operationalization of the concept of state failure, through a discussion of the indicators that are available, a consideration of what data would be desirable, and an exploration of some of the dynamics that can produce state failure. We begin therefore with a brief theoretical discussion of some of the conceptual choices that must be made in order to operationalize state failure. We then discuss some of the variables available to measure different conceptions of the phenomenon. Finally, we consider the possibility that state failure is a dynamic process difficult to capture in variable form. We construct a simulation to explore one particularly interesting dynamic of state failure in which governments attack the state apparatus, ultimately triggering the complete collapse of the state.

Theoretical Considerations

The intuition behind the idea of state failure is clear: there is a presumption that states ought to have certain capacities and behave in certain ways. When states do not meet these expectations, we perceive them as deviant in some way, and this deviance is captured by the use of the term "failure." However, we may not altogether agree on what states ought to do, and what capacities they ought to have. These differences can lead to disagreements about how to conceptualize state failure.

Thus there are various ways to approach the question of what constitutes state failure. The first choice among these possibilities hinges on how exceptional one considers state failure to be. Should state failure be treated as an extraordinary event, or is it a widespread condition?

In the first view state failure is a rare phenomenon. States are assumed to normally approximate the Weberian ideal-type, but under certain circumstances suffer catastrophic collapse. Following the Weberian model of the modern state, most analysts adopting this view emphasize the loss of the state's monopoly of the use of force as the key indicator of failure (for instance, Rotberg 2003, 2004). Loss of control over substantial portions of territory may also be taken as an indicator of failure.

This approach is essentially events-based. The failure of the state is treated as a more or less discrete event, with a fairly clear beginning and end. Research assuming this view of the phenomenon tends to focus on the antecedents of failure, to identify the circumstances that make it more or less likely. Much of this literature therefore treats state failure as a dichotomous variable (SFTF 2000, PITF 2003) or a relatively narrow set of categories (Rotberg 2003, 2004).

A difficulty with this view is that few states actually approximate their Weberian ideals even under normal circumstances (Callaghy 1987; Jackson 1990; Herbst 2000; Badie 2000). On most measures the states that experience failure-type events are not very different from "normal" less developed countries (Englehart 2007).

The events-based approach appears to hew closely to the concept of sovereignty, with failures representing those occasions in which states under international law may find their sovereign authority infringed, either by domestic rebel movements (which can seek foreign assistance if they control a substantial piece of territory) or by threats to the civilian population which the state cannot or will not address (leading to the possibility of humanitarian intervention). Yet the concept of sovereignty is itself notoriously problematic and difficult to operationalize (Jackson 1990; Krasner 1999, 2001).

A second view therefore is that states fail all the time, and the differences between them are matters of degree. In this view state failure is not an isolated event at all; it is continuous rather than discrete. The relevant questions then have to do with the degree of failure, the particular ways a state fails, and how failures change over time. From this perspective we can choose to examine a number of different ways in which states might fail. Loss of the monopoly of the use of force is one, but also the inability to police, collect taxes, prevent corruption or deliver services may also play a role.

Adopting this perspective may do some violence to the common perception that state failure represents a qualitative break from the ideal type of the Weberian state. However, it is important to recognize that ideal types are, as Weber says, "conceptually pure" mental constructs used to understand the relationships between phenomena generally, and therefore "not to be found anywhere in reality" (Weber 1949: 90). It is to be expected that actual, empirical states will deviate from the ideal type. Deviation is in fact the norm.

The recognition that all states deviate from the ideal allows us to address the conceptual problem that there actually seems to be a great deal of continuity between states that suffer catastrophic collapses and those that muddle on, failing continually without attracting significant attention from the international community. Research from this perspective tends to focus either on promoting good governance, e.g. by improving the delivery of services or reducing corruption (e.g. World Bank 1997). Alternatively it might focus on the effects of failure on citizens (Englehart 2009).

The difficulty with the continuous view of state failure is that it may not be sensitive to qualitative differences between states that muddle through and those that collapse catastrophically. The difference between these states in the existing data is often minimal. It may be that we need new data specifically targeted to the relevant theoretical concepts. It might also be that the difference has to do with strategic interactions between political actors which create differing dynamics, with some trajectories leading to complete collapse. This latter possibility is explored below by means of a simulation.

The decision about whether or not state failure is exceptional is fundamental. Once the basic choice between state failure as an event and as a condition is made, a host of subsidiary choices remain which bear on the operationalization on the concept. If state failure represents a loss of control by the central government, must it be total collapse, or can loss of control of substantial pieces of territory qualify? If state failure is continuous, what particular services must the state fail to deliver in order to qualify? Are only basic services such as security germane, or do others such as health and education qualify? In practice, these choices must often be made in light of the indicators available to capture them. We therefore turn now to the issue of operationalizing different conceptions of state failure and a discussion of the available data.

Operationalizing State Failure

There are few variables expressly constructed for the study of state failure, forcing researchers to either engage in costly data collection or rely on pre-existing off-the-shelf data. Here we look at a number of existing variables that may be used to operationalize different conceptions of state failure.

We divide these variables into four types. The first two groups treat state failure as a discrete phenomenon, appropriate to categorization. The latter two treat state failure as a continuous variable, essentially the inverse of state capacity. Because in this view all states fail to some degree, the data simply help us to assess the degree of failure in comparison to other states.[1]

1. Total collapse of governance

In this view state failure means the complete collapse of the central government. The classic examples would be Somalia since 1991 or Afghanistan from 1992 to 1996. Society typically decomposes into warring bands. Some try to assert themselves as the government of country, but are often prevented by coalitions of enemies following the logic of balance of power. Such situations developed in both Somalia and Afghanistan in the early 1990s. Control of the capital is generally regarded as the greatest prize, since this lends credence to claims to the international community to represent a legitimate government of the country, and thereby to secure foreign aid resources from the international community. However, other groups may try to secede (for instance, Somaliland) or to protect the autonomy of their local fiefdoms (as with many Afghan warlords after the 2001 war).

This understanding of state failure requires us to operationalize the concept as a dichotomous variable: either states have collapsed or they are functional. Although clearly important, such cases are relatively rare. One major difficulty is that this definition gives us a very small set of cases to work with. Extracting usable generalizations from such a limited set of cases can be quite difficult.[2] The major body of work to adopt this approach is that of the Political Instability Task Force (PITF, formerly the State Failure Task Force) (SFTF 2000, PITF 2003).

The Political Instability Task Force dealt with problems of a small universe of cases in two ways. One way was technical: it employed statistical techniques such as case-control methods and neural networks, designed to maximize the information available in a limited

[1] Note that this discussion excludes the proliferating multivariate indices of state failure, of which the most widely known is the Fund for Peace's (2008) Failed States index. These indices typically collapse a number of disparate indicators together with minimal (if any) theoretical accounts of why they should be included. Fund for Peace for instance includes twelve components, ranging from demography to the economy and politics. Yet no account is given of why these particular components were considered important, what other potential components might have been excluded, and why. This makes it difficult to know what exactly has been measured. The problem is compounded in the case of the Failed States Index by the fact that Fund for Peace supplies no information about how the measure was constructed and what source data was used.

[2] King and Zeng (2001a, 2001b) discuss the statistical issues involved in extracting maximal information from rare events.

number of cases. The other way was definitional: it adopted a broad definition of state failure that included not just the collapse of the central government, but also revolutions, ethnic conflicts, genocides, and all "adverse regime transitions," the latter meaning essentially any transition away from democratic rule.[3] This very broad definition of state failure maximizes the number of cases, but it does so at the risk of creating an incoherent dependent variable. It may be that these various phenomena are driven by quite different causal variables, making it difficult to interpret the results of an analysis combining them all in a single dependent variable. Indeed, the PITF definition of failure combines the total failure of all government institutions with ethnic violence that may or may not be instigated by the government, and with coups that may actually disturb very little of the functioning of the "everyday state." Elements of this definition may thus be at variance with the intuition behind the concept of state failure, which in this dichotomous conception should reserved for those cases where the institutions of government fail utterly.

One such measure is available in the Polity IV dataset, which employs an "interregnum" code for years when "there is a complete collapse of central political authority" (Marshall and Jaggers 2000, 16). This seems to capture the intuition behind the dichotomous conception of state failure more accurately than the SFTF/PITF variable.[4]

If we are willing to expand our definition of state failure beyond instances of total collapse, however, we may be able to do it in a way that is more internally consistent and does less violence to the intuition underlying the concept. Robert Rotberg, for instance, proposes a four-fold division of state failure based primarily on the state's control of violence (Rotberg 2004, 4-10). "Weak states" are states in which the government is unable or unwilling to deliver much in the way of public goods and services, and are characterized by high rates of corruption and crime. They have strong political, religious, communal and/or ethnic tensions with the potential to erupt into widespread violence – but have not yet done so. "Failing states" are states on the precipice, about to plunge into internecine violence. They are characterized by high violent crime rates, and much of the crime may be politically or communally motivated. "Failed states" are those that have plunged over the cliff into outright organized violence and civil war. Armed opposition groups challenge the central government, but at least a central government remains to challenge. "Collapsed states" are those in which the central government has collapsed entirely, and the country is ruled in effect by competing militias and armed groups, none of which is able to exert enough force to subjugate the others. This is the most extreme form of state failure. Rotberg includes a set of maps indicating which states he considered to fall into which category as of 2003. These indicate 28 weak states, 7 failing states, 6 failed states and only one collapsed state (Somalia).

For Rotberg, civil war thus distinguishes failed from merely weak states (Rotberg 2003, 5). There are a number of datasets that would provide dummy variables for civil war that could be employed for this purpose. The most commonly used is the Correlates of War (COW). An alternative is the PRIO dataset, which is more sensitive because of its lower cutoff for including a conflict (25 battle deaths per year or 100 for the entire conflict, as opposed to COW's cutoff of 1000 battle deaths).[5]

[3] One difficulty with this definition is that it is asymmetric: states are said to have failed when a democracy becomes undemocratic, but not when an undemocratic regime becomes a democracy. This means that democracies can fail in a way that non-democracies cannot, which may in part explain their finding that democracy is a risk factor for state failure (SFTF 2000).

[4] One issue with the Polity variable is that instances of collapse that result in foreign intervention may be coded as "interruptions", meaning foreign occupations, rather than state failures.

[5] A dummy variable for civil war is not, however, very information-rich. Civil war data indicating the degree of damage is also discussed below under section 4.

It is important to note, however, that the weakness of the state in other respects may increase the likelihood of armed conflict. Weak states make tempting targets for rebels (Fairbanks 2002; Fearon and Laitin 2003). Thus, if we define state failure as the presence of internal conflict, we may need to look to failures to provide security and provide other services, as discussed below in sections 3 and 4.

2. Collapse of state control over significant peripheral areas

It is common for states to lose control over peripheral areas. In such cases the government typically retains control of the capital, but rebels or local bosses effectively control other parts of the country – sometimes very large sections. For example, prior to the collapse of the Somali state in 1991, President Siad Barre was sometimes derisively referred to as the "mayor of Mogadishu." Afghan opposition figures have begun to refer to Hamid Karzai similarly as the "mayor of Kabul". Such cases are much more common than total collapse, including for instance Burma, Congo, Sierra Leone and Lebanon.

The SFTF phase IV data allows us to segregate failed states in the problem set, on the basis of the degree of control over territory. It codes adverse regime changes into four categories: *Magnitude 1* indicates a coup or other democratic setback with little impact on the government's control over territory; *Magnitude 2* indicates the state has lost control over a limited area; *Magnitude 3* indicates a failure of state authority in a substantial part of the country and/or the capital; *Magnitude 4* is reserved for complete collapse or near total failure. Since this variable is limited to "adverse regime changes," it eliminates civil/ethnic wars and genocides/politicides unless accompanied by a breakdown of the state apparatus.

Polity IV contains a similar variable, called "fragmentation." This indicates whether some significant portion of a country's territory is not controlled by the central government, on a scale ranging from 0 (no fragmentation) to 3 (25-50% of the country's territory not under the control of the central government). Governments that control less than 50% of the territory are coded as interruptions or interregnums, as discussed above.

This definition of state failure has the benefit of being readily observable. One difficulty is that it suggests that the failure of the state is simply territorial. Yet such states typically fail in other ways that are relevant to their loss of control over territory. Extensive corruption might create conditions that lead to the loss of territorial control, for instance by alienating the support of local populations, eroding the state's ability to extract resources, or even enabling local military officers to sell arms to insurgent groups. Like civil war, the loss of territory in fact may be only one symptom of deeper underlying structural failures of the state. This leads us to our next two possible definitions.

3. Inability to police territory, enforce the law, or deter corruption

It is universally agreed that, at a minimum, states ought to provide basic public goods such as law enforcement, security for citizens, the enforcement of contracts, as well as preventing corruption and abuse of power by their employees. Yet many states have great difficulty providing these goods, and this affords us one way of measuring state failure continuously. One could think of measuring state failure in this way as testing the degree to which states deviate from one of two ideal-typical standards: the monopoly of the use of force, and the sovereign authority to make and enforce laws with a territory.

Failures to control violence, provide policing and enforce the law are common to all states to some degree, so it is reasonable to begin thinking of state failure as a continuous rather than categorical variable when adopting this approach. Indeed, these problems are so widespread that we can include all states in the data, asking not whether they have failed but to what degree each fails. The issue then is to identify indicators that will help us compare all states with respect to their degree of failure.

Perhaps one of the most obvious indicators of a loss of control over the use of violence is the presence of armed insurrection. While the mere presence of civil war may, as indicated above, be used as a measure of state failure, a more nuanced indicator would be the severity of civil war. One measure of severity is battle deaths. Gleditsch (2004) has constructed data on battle deaths using revised Correlates of War data. This data has a relatively high cutoff of 1000 battle deaths to be included. Somewhat more sensitive is Lacina and Gleditsch's (2005) data, based on the PRIO civil wars list, with a cutoff of 25 battle deaths. Battle deaths may, of course, be a crude measure of the degree of disruption caused by civil violence. Marshall has constructed a dataset that incorporates factors such as civilian casualties, refugee flows, infrastructure damage, war-related disease, loss of confidence in government, and so on, to assess the overall impact of war (Marshall and Gurr 2005).

One could also measure the degree to which a state can maintain law and order within its territory. Crime statistics might initially seem attractive for this purpose, but the existing data is actually quite poor. Interpol collects data from member states, but is explicit that the data cannot be used to make cross-country comparisons, because crimes are reported differently in different countries. Slightly better is the World Health Organization's data on homicides, which is collected using standardized diagnostic codes and is therefore more comparable across countries. However, this data is collected only from the 70 or so countries that require a death certificate. Furthermore, it is probably skewed: it will be more accurate for countries with better record-keeping, which are likely to be the higher-capacity states. Deaths in these places are more likely to be recorded, thereby giving the appearance of more homicides than in places where more deaths go unrecorded.

The Political Risk Service (PRS) is a professional risk-assessment firm which produces a number of useful indicators for sale to their clients. PRS constructs a law and order variable based on surveys of country experts, which measures the strength and impartiality of the legal system and the degree to which the law is generally respected and order is observed by the general population. While this is not an objective indicator, it is less prone to the systematic bias inherent in crime statistics.[6]

One of the better sets of state capacity indicators are those developed by Kaufman and his collaborators (1999, 2002, 2003). Based on an unobserved components analysis of a variety of indicators, this data is arguably superior because the combination of sources makes it more reliable than any of the component indicators. It is available for a large number of countries, but for only four years (1996, 1998, 2000, 2002). Included in this data is a rule of law measure similar to the PRS law and order variable.

One possible measure of the degree to which states can constrain the behavior of their own employees is Transparency International's Corruption Perceptions Index (CPI). This too is based on surveys of experts, and has the limitation of including a relatively small number of countries. Kaufman et al. include a corruption variable as well, which (like their law and order variable) covers a large number of countries but relatively few years.

Infrastructure such as roads, railroads and telephone main lines may also provide a useful index of central government control of peripheral areas (Herbst 2000). Measures for all three are found in the World Bank's *World Development Indicators*, but coverage is uneven and probably skewed towards higher capacity states, which are more likely to report such data.

Resource extraction can serve as a general indicator of state capacity/failure, with tax as a proportion of gross domestic product probably the best measure (Arbetman and Kugler

[6] It is, however, important to note Evans' complaint that such ratings amount to "beauty contests" which better capture the impression states make on business and academics rather than actual underlying conditions (Evans, 2005: 39).

1997, Fauvelle-Aymar 1999, Lieberman 2002). Extracting taxes has traditionally been regarded as one of the most important functions of the state. Indeed, in historical sociology it is generally assumed to drive the process of state formation, which is sometimes depicted as almost entirely about collecting revenue to pay for a powerful military (Hintze 1975; Tilly 1975, 1992; Finer 1975; Brewer 1989). As Tilly puts it, "preparation for war ... involves rulers ineluctably in extraction. It builds up an infrastructure of taxation, supply, and administration that requires maintenance of itself and often grows faster than the armies and navies that it serves" (1992, 20).

The state's ability to extract taxes is thus at the heart of its control over the use of violence, or indeed of its ability to engage in any other policy. In order to collect taxes effectively states need to be able to collect information about their citizens, induce them to pay, sanction defaulters, and keep tax officials reasonably honest. States that are better able to collect taxes therefore should be better at policing their territories. In turn this should lead to better control over the use of violence.[7]

Fearon and Laitin (2003) interpret GDP per capita as a measure of state capacity, and so inversely it also might be used as an indicator of state failure. Their logic is that GDP reflects the resources available to the state to invest in police and the military, and therefore of the state's superiority over insurgent groups. Following this logic, however, it seems that tax as a proportion of GDP would be an even better indicator, since it reflects not just the potential pool of resources available, but rather the actual resources tapped.

Revenue is another possible indicator of the resources available to the state, but it is problematic. Revenue derived from non-tax sources may neither indicate state capacity nor foster it. For instance, revenue from foreign aid represents the negotiating skills of a country's leaders more than any capacity to use funds productively. The literature on rentier states suggests that revenue from natural resource endowments may actually inhibit the development of state capacity and promote corruption (Mahdavy 1970; Beblawi and Luciani 1987; Vandewalle 1997; Rubin 2002).

Size of military and police forces might also be regarded as good indicators of the strength of the state. They actually turn out to be problematic, because they say nothing about the competence and effectiveness of those forces, or the degree to which they are actually under the control of the government. Since agency loss is a serious problem in failing states (Englehart 2009), a large military may actually be a significant hazard when it is not effectively under the control of the government.

4. Inability to provide services such as education, health care

One could also define state failure as the failure to deliver public goods expected by citizens. This is, of course, extremely common. As with the inability to police, it would be best to

[7] Using tax as a proportion of GDP as a measure of state capacity does assume that states ordinarily collect as much revenue as they can, given the institutional constraints they face. The implication of this assumption is that tax policy, as much as it might be a hot-button political issue in many places, actually makes little difference to tax receipts. This may seem a strong assumption, but the data suggests it is probably accurate. For instance, after Ronald Reagan's highly touted tax reform in 1986, tax collections in the United States actually went up, in both absolute terms and as a proportion of GDP. Bueno De Mesquita et al. (2005) argue that parliamentary governments have a higher propensity to tax than presidential systems, but overlook the possibility that this may be because they are on balance higher capacity. Parliamentary systems do have higher tax collections as a proportion of GDP than presidential systems, but they also score better on all other measures of state capacity, including corruption and law and order.

consider these indicators of failure as continuous variables, which help us assess the degree to which each state fails in its ideal-typical duties.

One difficulty with this approach is that it requires making fairly strong assumptions about what states ought to do. This may raise normative issues about what constitutes good policy. Furthermore, it is not necessarily constant: what we expect from states has increased over time (Thomas et al. 1987). For instance, the notion that states are responsible for the education and health of their citizens is fairly recent. Similarly, prior to the twentieth century few people believed that the state could have a significant and positive impact on the economy. States were thus not responsible for creating stable macroeconomic conditions that would lead to growth. This potentially affects measures of state capacity/failure: a state that performs well by the standards of 1900 would be doing very poorly by the standards of 2008.

Variables that could be employed for such a measure include literacy rates, the provision of schooling, health care delivery and so on. Many of these measures are available from the World Bank's *World Development Indicators.* For many purposes, the Bank's Human Development Index is a useful measure.

What State Failure is Not

A number of confusions have crept into the literature on state failure. The most common involve failing to distinguish governments from states, and with confusing policy outputs from the conditions that are likely to generate them.

State failure is sometimes confused with regime stability or longevity. This is problematic, because some regimes actually depend on the weakness of the state apparatus to ensure their survival. For instance, a government worried about the loyalty of the professional military may create private militias as a counterweight. These militias are associated with the government but not the state, undermining the state's monopoly of violence. Aristide's adoption of the *chimeres* in Haiti, for instance, was through his political party rather than through the state apparatus. This meant that it was responsive to Aristide personally and could serve as a check on the police forces, which he mistrusted. In Iraq, Saddam Hussein created the Fedayeen Saddam as a similar private militia controlled by his sons because he suspected coup attempts from the regular military and the Republican Guard.

Regimes may also rely of the weakness and corruption of the civil service to enhance their own patronage systems (Chabal and Daloz 1999). This enables leaders to construct patrimonial systems which are personally profitable and can ensure long tenures with a high degree of stability, but which require that the civil service be unable to compete as a provider of public goods, including law and order. Disorder "offers opportunities for those who know how to work that system" (Chabal and Daloz 1999, xix).

A related issue is democratic accountability. It is sometimes assumed that democratic governments preside over stronger states because they institutionalize higher levels of accountability. There does in fact seem to be a relationship between democracy and state capacity, but it is also true that many democratic governments preside over remarkably weak states. Furthermore, the fact of a transition to or from democracy generally has little short-term impact on measures of state capacity. In Colombia, for instance, a significant democratic opening in the early 1990s has had little impact on the capacity of the state to control non-state armed groups such as paramilitaries and insurgents. Although there may be a long-term relationship between democracy and state capacity, for most purposes it is safer to treat the concepts as distinct.

For similar reasons, it is also probably better to treat state failure and state capacity as distinct from particular policies enacted by governments. Most notorious in this regard are the anti-corruption and structural adjustment programs sponsored by the international financial institutions in the 1980s and 1990s. Although touted as improving government performance by shrinking unnecessary bureaucracy and fighting corruption, in many countries they served as an excuse to fire competent civil servants who are not politically reliable, or cut programs that were not useful for patronage purposes. Rather than improving bureaucratic performance, therefore, such programs could significantly degrade it.

In order to assess the threat of state failure accurately, we will need to develop new data more closely tailored to the empirical questions at hand. Off-the-shelf data can only go so far. The specific types of data we will need will be tied to the specific forms of state failure and the specific causal chains that lead to them.

In addition, we need to develop more sophisticated accounts of the dynamics of state failure. As noted above, failing states typically do not differ greatly from non-failing less-developed countries. This could be due to some unobserved component that could theoretically be captured as a variable. However, it might also be due to political dynamics and strategic interactions that are not easily captured as variables. Carment (2003) for instance notes the need to integrate long- and medium-term factors such as economic development and the presence (or absence) of conflict with the short term dynamics of political decision-making. We need to develop strategies to identify such dynamics.

One such strategy is to examine the dynamics of the interaction between variables through simulation. An advantage of simulation is that it allows us to model and analyze these causal chains so that we can begin to understand which factors are most important in driving them. It is especially useful when we are studying a phenomenon where many conflicting influences are present, and where the choices of decision makers at particular times during the process may be crucial in producing different outcomes. Simulation also allows us to take substitutability seriously (Most and Starr 1989) and examine several different means that governments can use to pursue the same goal.

Simulating State Failure and Collapse

The simulation that we use in this chapter was adapted from one used to analyze state capacity in the context of "two-level" security threats (Simon and Starr 1996, 2000, 2002). It is derived from a model which describes how governments react to the simultaneous possibility of both international and domestic threats to security. It poses two basic policies that governments use to address threats: extraction of resources (from society or allies) and allocation of resources (to buy off or co-opt threats). We make minor modifications to the model in order to analyze the process, described below, where weak states face a primarily domestic threat but must rely on extraction, allocation, or the creation of additional non-state militia groups in order to maintain the viability of government.

Governments against States: A Pattern of State Failure

As an illustration, we discuss here one pathway to state collapse. This is only one course to state failure; many others are possible. It does seem to be an especially important one, however, since it is implicated in the two most complete collapses of the post-Cold War world: Afghanistan and Somalia.

In this trajectory, it is governments that destroy states (Englehart 2007). This may seem like an odd formulation, since in studies of the less developed world it is unusual to draw the distinction between state and government. In the study of the advanced industrial countries, however, it would be a basic error to confuse the government with the state apparatus – the civil service and the professional military and police forces. One need not be a fan of *Yes, Minister* to understand that the state apparatus may have very different background, preferences and incentives than the elected government.

The same is true in less developed countries, where the government—elected or not—may have very different education, interests and incentives from the civil service and military. Indeed, the civil service and military can be a threat and potential site of resistance to governments. As a consequence, governments have a strong interest in controlling the threat posed by the state. This can be accomplished in a variety of ways: the promotion of loyalists, the appointment of political hacks or military officers to key positions, cutting off resources, the liberal use of intimidation and bribery, or some combination of these. At the most extreme, governments may actually attack the state apparatus through purges of the civil service and military, or by setting up militias as a counter-weight to the police and military.

This path to collapse assumes that the rise and fall of state capacity is a continuous rather than discrete process and that governments may be tempted to adopt policies that help their short-term survival but this may lead to unintended collapse in the long-run.

But what are these short-sighted policies, and under what conditions might they lead to collapse? The object of our simulation is to explore these questions. If fruitful, this investigation will provide us clues about the dynamics of real world state failure and collapse.

Simulation Description and Analysis

We assume that governments face an array of external and internal threats, and that they use capabilities to address these threats. According to our two-level security model (Simon and Starr 2000), governments perceive the level of threat based on a combination of three factors: interaction opportunity, willingness to attack, and strength or capabilities of the opponent. For internal threats, willingness is the government's estimate of the extent of change desired by each rebel group, on a continuum from revolution to reform – not the group's willingness to attack.

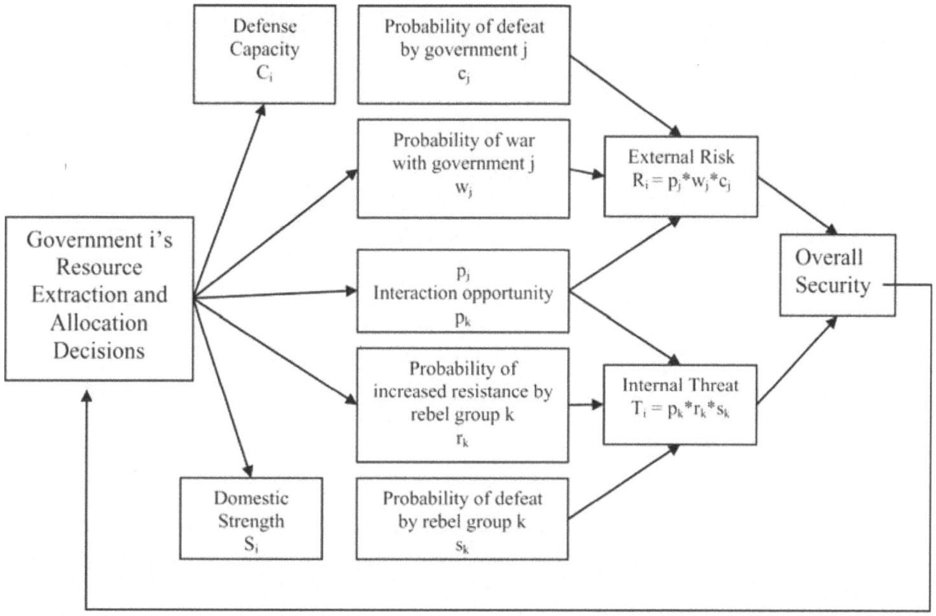

FIGURE 1 Summary of Functional Relationships in the Simulation Model.

To evaluate its overall security, the government sums up all of its threats and conceptually compares this to its current level of capabilities. It wants to avoid the situation where threats are greater than capabilities, and so it will take action to improve this ratio. First, we assume that given limited time and attention, the government will, in each "round" of the simulation, try to reduce its most pressing threat. This fits the realist logic that governments seek to "minimize their maximum threat."[8] Our earlier work on two-level security focused on the trade-off that must be made to address domestic and international threats. Efforts to promote security at one level might lead to vulnerability on the other. However in the current paper, we are concerned about a different trade-off—between security and state capacity.

Government Actions

In the simulation, a government can use two basic means to address threats: extraction to build capabilities (for deterrence, or intimidation), and allocation of resources to bribe opponents and reduce their willingness to attack.

Extraction comes from three possible sources: societal resources, allies, and as a transfer of resources between those directed at internal versus external threats. Governments have access to a pool of societal resources, and they possess an extraction capacity—analogous to the ability to tax—that is based on the amount of coercive resources controlled by the state. Following logic outlined above, when a government extracts (or is given aid) from an ally, this provides usable resources but does not increase the state's capacity to extract (see Appendix 1 for formulas and initial values for all simulation parameters).

Using allocation, a government can reduce the threat it faces from a rebel group by outright bribery, and also through the provision of other resources to the group or its supporters. The simulation assumes that resources spent in this fashion will reduce the willingness of the rebels to violently oppose the government—moving them more toward the reform end of the scale. In our earlier work this was considered the more dovish of the two strategies for addressing threat.

A third method to reduce threat, which we categorize as a form of extraction to build capabilities, is for the government to establish militias outside the state, or fund new rebel groups with the hope that they will divide and distract the main opponents of the regime. These government-created groups are modeled like other rebels, but with a 'willingness to resist' value near zero, so that they are not (at first) threatening to the state.

In each round of the simulation (Figure 2), the government assesses the threats it faces, and chooses to extract or allocate based on its internal political disposition (see Ikle 1971) as well as learning from the results of 5 prior rounds (each government is assigned a probability of extraction in any given round to reflect its disposition, and the 'decision' is made by a random variable). The government then extracts or allocates a standard amount of resources (about 3% of its base) to address its most pressing threat. This process occurs simultaneously for all states in the system (we use 10 states as an approximation of a regional system). After the round, capabilities and threats are recalculated and then the next round begins.

[8] While neither the classic balance of power literature nor Walt (1987) specifically call for the strategy of minimizing the maximum threat, this strategy is implicit in these formulations. Moreover, this strategy also minimizes decision costs for actors, consistent with bureaucratic politics and cybernetic models (Steinbrunner 1974; Allison 1971) which motivate our assumption that decision makers act under bounded rationality.

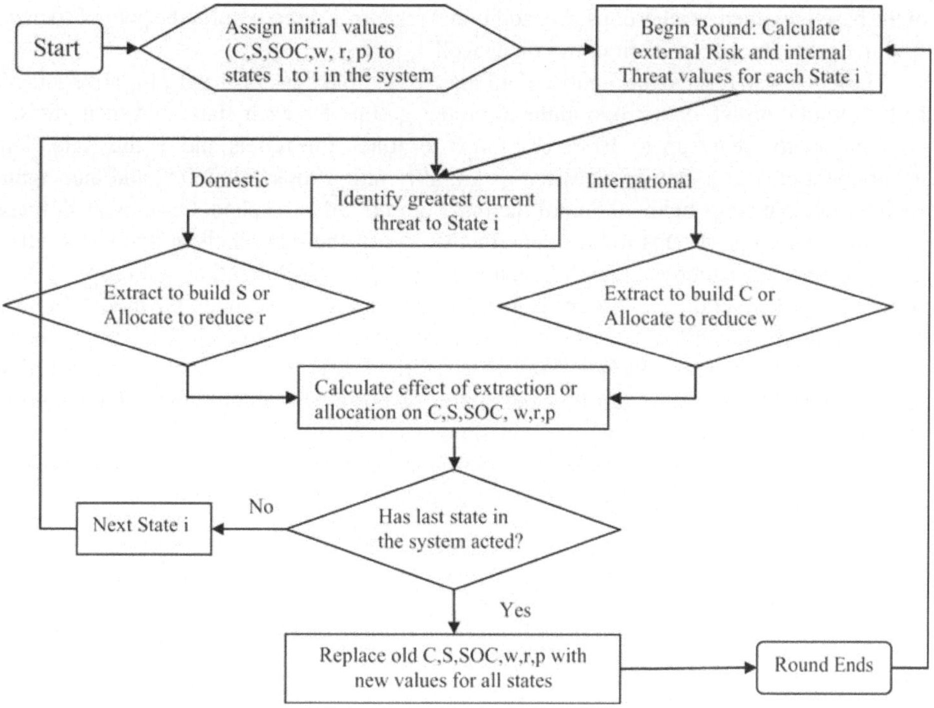

FIGURE 2 Basic Simulation Flowchart.

Rebel Groups

To model the processes that can lead to state failure, the reaction of rebel groups to the government policies and to each other needs more attention. Much of the rebel group activity in the simulation is a reaction to government policies. Rational actor analysts (Lichbach 1995; DeNardo 1985), political process theorists (Tilly 1978), and many empirical studies (Tilly 1975; McAdam 1996) stress that government policies and levels of repression affect the resources, mobilization ability, goals, and success of opposition groups. In our simulation, the government's extraction and allocation decisions are the main mechanism affecting the opposition group's strength and willingness to intensify their resistance.[9] After a round in which the government extracts, each rebel group's strength grows by about 1%. This is based on an assumption—consonant with rational actor, resource mobilization, and political opportunity models of rebellion—that rebels grow based partly on their current endowment of resources.

In addition, empirical examples in failing states suggest that rebel groups can gain resources through forms of taxation or through the procurement of lootable resources (Reno 1999; Snyder 2006). We assume that this is most likely when the state's capacity is weak. In the simulation, when a state falls below a threshold of extraction capacity (arbitrarily, 50%

[9] Lichbach (1995) outlines over two dozen strategies used to overcome the free rider problem to create conditions for the growth of rebellion. In his terms (1995:22-26), improvements in strength ratios help the rebels mobilize by increasing the probability of winning (1995:62); extraction decisions help reduce the size of the public good (1995:62); allocation decisions increase resources (1995:48), increase benefits (1995:35), and lower costs (1995:36) to those who might join the opposition.

of the baseline), then rebel groups gain additional resources directly form the pool of societal resources that the government draws on as well.

The rebels also react to the number of militias/rebel groups established by the government. In the simulation we create two authentic rebel groups for each state, and then allow a government to create up to 10 other groups to attack the rebels and/or the state. The government creates a new group when its viability ratio drops below 0.5, and then again each 5 rounds it stays below 0.5 until the maximum of 10 is reached. Once created, these government-created groups reduce the effectiveness of the real rebels in two ways. First, they compete for resources, thus for each rebel group the mobilization rate drops by 1%. Second, they distract the real groups from dealing with the government. We model this as a reduction their interaction opportunities with the government (p). Applying an argument from Deutsch and Singer (1964) with regard to multipolarity, interaction falls as a proportion of rebel-government dyads, up to a maximum of 66%. This effect allows these militia groups to dramatically improve the government's security. However, they are costly in terms of resources. The government must continue to fund its groups in order to keep them on their side; they get 10% of what the government extracts from society.

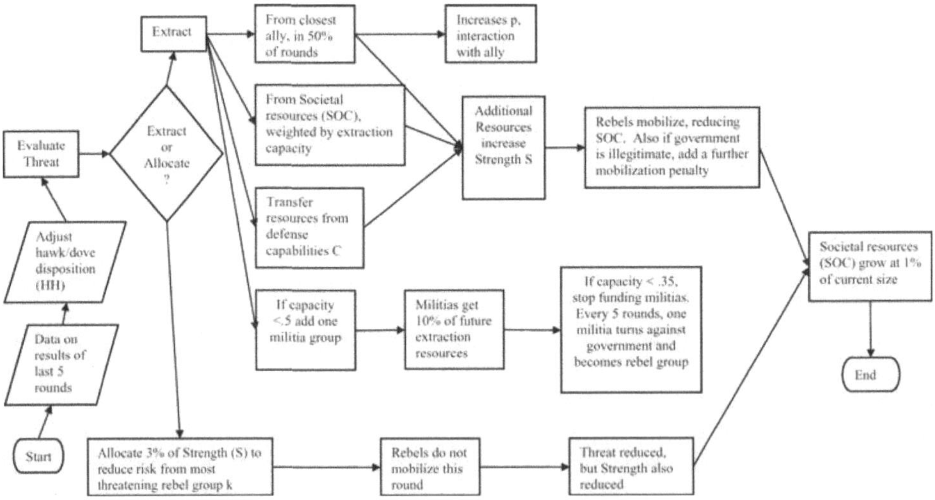

FIGURE 3 Effects of Government Response to Internal Threat

The algorithm described above is illustrated in Figure 3. Where possible, we chose values for the simulation parameters (Appendix 1) that would make sensible approximations to real world data. Growth in societal resources, which is analogous to GDP growth, averages 1% per round. The number of initial rebel groups (2), is an approximation of the number of real or potential oppositions that most state face (Gurr and Scarritt 1989). One's natural skepticism about computer simulation—that it produces "made-up data" and is easily manipulated to produce what researchers want to show—is tempered by actually constructing, debugging, and using this program. It turns out that the complexity of the algorithm and parameter relationships—which is all based on theoretically grounded logic—makes it very difficult if not impossible to manufacture results without using ridiculous parameter values. We believe that the simulation provides a useful tool for analyzing the relative importance of factors which affect state failure.

Simulation Results

Figures 4a and 4b illustrate the path to state collapse that we outlined above. They portray the decline of a weak state that, under baseline conditions, would have remained weak but survived over time, with its state extraction capacity near the baseline value of 50 over the long term. However, we forced this government to fall into the trap of adding militia groups when extraction capacity fell below 50 (our arbitrary threshold) near round 35. At first, these militia groups provided great relief from internal threats, as evidenced by the sharp fall in threat through round 66 (Figure 4a). Threat remains stable and low through round 90.

Unfortunately, however, state extraction capacity continued to fall during this period, because of the need to fund the militias. Societal resources also fall because of the presence of the militias. In Figure 4b, note that the government's strength to threat ratio continues to decline between rounds 66-90, though threat is constant and low. Once capacity drops below 35 (our second arbitrary threshold), the lack of both capacity and societal resources require the government to cease support of its militias. Beginning in round 90, one by one, these militias turn against the government, escalating the threat level each time. The government simply did not have the resources or extraction capacity to cope with this increase, as shown by the drop in the strength to threat ratio in Figure 4b. The government tries to extract what it can to fight the increasing threat, but low extraction capacity means it cannot gain strength fast enough to counter the threat of more numerous rebel groups. As the decline continues, societal resources are completely depleted by round 110, making recovery impossible.

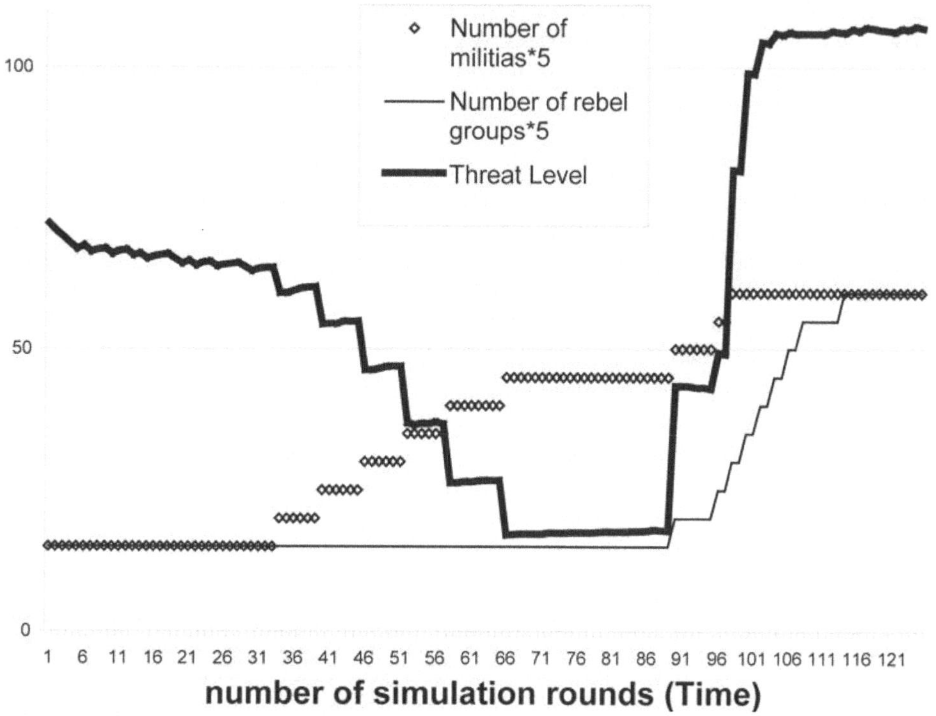

FIGURE 4a One Path to State Failure

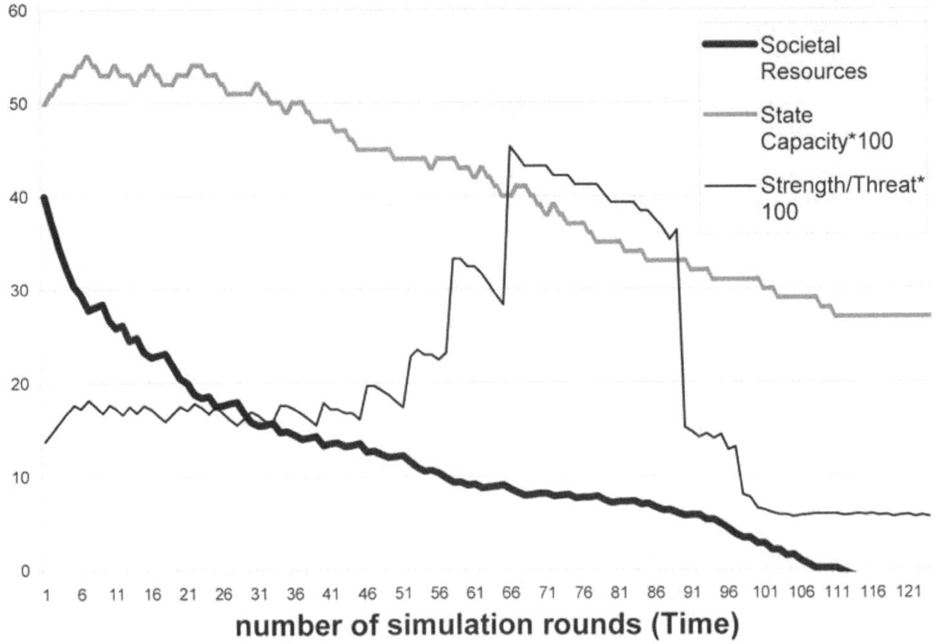

FIGURE 4b One Path to State Failure

Interestingly, the simulation shows that the collapse of societal resources precedes the final decline in state capacity, indicating that resource-rich states may be able to hold off state failure for a long time, despite their weak capacity. To further explore such patterns, we varied simulation parameters and compiled data on average outcomes of multiple runs of the simulation in Table 1.

As a baseline, we created a system of 10 interacting states with parameter values similar to those in our previous work (Simon and Starr 1996, 2000). We ran this system for 150 rounds and then repeat this for 100 iterations. The results in Table 1 are average changes across all 100 iterations (this is necessary because government allocation/extraction strategy is chosen by a random variable, and the randomness of the choice can lead to divergent outcomes). We then varied some of these parameters by roughly + or - 50% and try to gauge their relative impact on security (S/T) and state capacity.

Let us begin with some impressionistic results. The system modeled in the simulation was surprisingly stable. Governments are constantly trying to improve their security ratios, and they also adjust their strategies (extraction or allocation) based on the pattern of success or failure in the last 5 rounds. Generally, when faced with shocks to their security levels, even very weak governments are able to recover to their baseline levels in 20-30 rounds. The simulation, at its baseline values, tends toward equilibrium. This makes sense given the rarity of actual catastrophic collapses.

Many of the factors affecting state failure are not surprising. The ability to extract ally resources—even though these resources did not contribute directly to the state's extraction capacity—insured state survival. The ability to continue to build societal resources, despite the severity of any external or internal threat, is the most crucial predictor of success.

From Table 1, we see that a smaller system size, higher initial state capacity, higher extraction rate, higher legitimacy, and more frequent ally extraction, all helped the failing state improve its security and capacity outcomes relative to the baseline case. Of these, legitimacy, extraction rate, and ally extraction (rows H, K, and M) had the strongest effect

TABLE 1 Effects of Causal Factors* on State Security and State Capacity

SYSTEM	International security C/R change (%)	Domestic security S/T change (%)	Societal resources change (%)	% Change in Extraction frequency (international threats, H)	% Change in Extraction frequency (domestic threats, HH)	State Capacity change (%)	Average Number of militias/ rebel groups at simulation end
A Baseline system** Major power average results:	2.6%	2%	113%	-3.4%	-5%	-22%	0/0
B Baseline system, Minor power average results:	-19%	9%	132%	-5.6%	-24.6%	-12%	0/0
C Baseline system, Failing State (FS)†	-42%	-38%	-91%	22%	16%	-50%	2.9/6.9
D Larger System size: FS in 14 state system	-54%	-4.8%	-99%	31.2%	20.6%	-52%	1.9/7.9
E Smaller system size: FS in 6 state system	-32%	130%	-66%	11%	6%	-46%	4.2/5.6
F FS, Higher initial state extraction capacity (.75),	-6.5%	8.4%	-92%	14%	2%	-40%	2.8/1
G FS, Lower initial state extraction capacity (.25) FS	-10.3%	-6.6%	-99%	39.6%	24%	-60%	0/10
H FS, with extraction rate always (HH) > 75%	-5%	41%	-98%	7.2%	8.3%	-30%	3.3/6.3
I FS, with extraction rate always (HH) < 25%	-6.9%	-8.4%	-75%	-40.4%	-19.2%	-64%	2.2/7.0
J FS, Lower legitimacy (G=0.02)	-9.5%	-4.8%	-99%	38%	27.2%	-58%	1.1/8.8
K FS, Higher legitimacy (G = -0.05)	-0.7%	43%	-10%	-5.2%	-6.6%	-34%	7.3/1.8
L FS, with ally extraction in 10% of rounds	-6.9%	9.6%	-83%	18.4%	10%	-46%	3.4/6.4
M FS, with ally extraction in 50% of rounds	-4.2%	38%	-15%	1.5%	-5%	-30%	6.9/1.8

*Shows effects of System Size, State Capacity, Extraction/Allocation Policy, Legitimacy, and Ally support.

**Baseline system has 10 states, 2 major powers, 8 minor powers, 2 rebel groups per state. The simulation runs for 150 rounds. Averages shown are for 100 iterations of the simulation. This is necessary since there can be variation in each simulation run due to random variables.

†Failing State baseline values: Capabilities=Strength=10; Societal resources =50; State extraction capacity = 0.5; extraction frequency (HH) = 50%; Legitimacy G=0.01; Ally extraction in 0% of rounds.

in helping weak states survive. The reason is that these factors allowed the failing state to conserve more of its societal resource base, especially early in the simulation; this helped reduce the decline in state capacity later in the simulation run.

The size of the regional state system in which the failing state finds itself had a lesser but interesting effect on state extraction capacity. A smaller system created less external threat for the failing state, and thus a higher C/R ratio to start. The lack of external threat allowed the government greater use of its external security resources for internal security, creating a better internal security (S/T) ratio. However, by the end of the simulation run its capacity dropped almost as badly as in the baseline case, because it still depleted societal resources at the same overall rate.

The choice of strategy—extraction to build capabilities and deter, versus allocation to buy off threats—also showed an interesting and unexpected pattern. Governments optimize this strategy based on the security results in the last 5 rounds. Our results show that states nearing collapse (such as the low legitimacy case in row J) try to increase their extraction rate. But this comes too late to save them. When forced to extract at a high rate (75% or more) earlier in the simulation, the state had more success (row H). This is because of the extra extractions occurred when state extraction capacity and the societal resource base was higher, and thus more resources were obtained from these extractions than from those that occurred later in the simulation when extraction capacity and resources were nearly depleted. This evidence supports the conclusion that preservation of societal resources is most crucial for helping a failing state to survive.

Conclusion

While the simulation results are not startling, they do represent a systematic attempt to derive implications from the particular path to state failure and collapse developed in the first part of the chapter. The simulation also provides us with a tool for further exploration of the sets of factors which affect state failure, especially those related to the regional subsystem in which the endangered state is located. Finally, the simulation reaffirmed the empirical observation that it is difficult for states to actually collapse—much is built into the system to allow a government to survive, even with low resources and high threat from rebel groups.

References

Allison, G., 1971. *Essence of Decision*. New York, NY: Harper Collins.

Arbetman, Marina and Jacek Kugler, 1997. *Political Capacity and Economic Behavior*. Boulder: Westview.

Badie, Bertrand, 2000. *The Imported State: The Westernization of the Political Order*. Stanford, CA: Stanford University Press.

Beblawi, Hazem and Giacomo Luciani, eds., 1987. *Nation, State and Integration in the Arab World: Volume II, The Rentier State.* NY: Croom Helm.

Brewer, John, 1989. *The Sinews of Power: War, Money and the English State, 1688-1783.* New York, NY: Knopf.

Bueno De Mesquita, Bruce, George W. Downs, Alastair Smith and Feryal Marie Cherif, 2005. "Thinking Inside the Box: A Closer Look at Democracy and Human Rights," *International Studies Quarterly* 49:3, 439-58.

Callaghy, Thomas M., 1987. "The State as Lame Leviathan: The Patrimonial Administrative State in Africa," *The African State in Transition*, Zaki Ergas (ed.). New York, NY: St. Martin's Press, 87-116.

Carment, David, 2003. "Assessing State Failure: Implications for Theory and Policy", *Third World Quarterly,* 24:3, 407-427.

Chabal, Patrick and Jean-Pascal Daloz, 1999. *Africa Works: Disorder as Political Instrument.* Bloomington: Indiana University Press.

DeNardo, J., 1985. *Power in Numbers: The Political Strategy of Protest and Rebellion.* Princeton, NJ: Princeton University Press.

Deutsch, Karl and JD Singer, 1964. "Multipolar Systems and International Stability," *World Politics,* 16:3, 390-406.

Englehart, Neil A., 2007. "Governments Against States: The Logic of Self-Destructive Despotism," *International Political Science Review,* 28:2, 133-153.

——, 2009. "State Capacity, State Failure and Human Rights," Forthcoming in *Journal of Peace Research,* 49:2.

Evans, Peter, 2005. "Harnessing the State: Rebalancing Strategies for Monitoring and Motivation," *States and Development: Historical Antecedents of Stagnation and Advance,* Matthew Lange and Dietrich Rueschemeyer (eds.). NY: Palgrave Macmillan, 26-47.

Fairbanks, Charles H. Jr., 2002. "Weak States and Private Armies," *Beyond State Failure? Post-Colonial Africa and Post-Soviet Eurasia in Comparative Perspective*, Mark Beissinger & Crawfurd Young (eds.). Washington, D.C.: Woodrow Wilson Center, 129-160.

Fauvelle-Aymar, Christine, 1999. "Political and Tax Capacity in Developing Countries," *Kyklos,* 52:3, 391-414.

Fearon, James D. and David D. Laitin, 2003. "Ethnicity, Insurgency and Civil War," *American Political Science Review,* 97:1, 75-90.

Finer, Samuel, 1975. "State- and Nation-Building in Europe: The Role of the Military," *The Formation of National States in Western Europe*, C. Tilly (ed.). Princeton: Princeton University Press, 84-108.

Fund for Peace, 2008. Failed States Index. <http://fundforpeace.org>. Accessed July 25, 2008.

Gleditsch, Kristian Skrede, 2004. "A Revised List of Wars Between and Within Independent States, 1816-2002," *International Interactions,* 30, 231-262.

Gurr, T.R. and J. Scarritt, 1989. "Minorities at Risk: A Global Survey," *Human Rights Quarterly,* 11, 375-405.

Hintze, Otto, 1975. *The Historical Essays of Otto Hintze,* Felix Gilbert (ed.). New York, NY: Oxford.

Herbst, Jeffrey, 2000. *States and Power in Africa: Comparative Lessons in Authority and Control.* Princeton, NJ: Princeton University Press.

Ikle, F. C., 1971. *Every War Must End.* New York, NY: Columbia University Press.

Jackson, Robert, 1990. *Quasi-States: Sovereignty, International Relations and the Third World.* New York, NY: Cambridge University Press.

Kaufmann, Daniel, Aart Kraay and Pablo Zoido-Lobatón, 1999. *Governance Matters,* World Bank Policy Research Paper 2196.

——, 2002. *Governance Matters II: Updated Indicators for 2000-01*, World Bank Policy Research Paper 2772.

Kaufmann, Daniel, Aart Kraay and Massimo Mastruzzi, 2003. *Governance Matters III: Governance Indicators for 1996-2002*, World Bank.

King, Gary and Langche Zeng, 2001a. "Explaining Rare Events in International Relations," *International Organization,* 55:3, 693-715.

——, 2001b. "Improving Forecasts of State Failure," *World Politics,* 53, 623-58.

Krasner, Stephen, 1999. *Sovereignty: Organized Hypocrisy*. Princeton, NJ: Princeton University Press.

——, 2001. *Problematic Sovereignty.* Princeton, NJ: Princeton University Press.

Lacina, Bethany and Nils Petter Gleditsch, 2005. "Monitoring Trends in Global Combat: A New Dataset of Battle Deaths," *European Journal of Population,* 21:2, 145–166.

Lichbach, Mark I., 1995. *The Rebel's Dilemma*. Ann Arbor: University of Michigan Press.

Lieberman, Evan S., 2002. "Taxation Data as Indicators of State-Society Relations: Possibilities and Pitfalls in Cross-National Research," *Studies in Comparative International Development,* 36:4, 89-115.

McAdam, D., 1996. "Conceptual Origins, Current Problems, Future Directions," *Comparative Perspectives on Social Movements,* D. McAdam, J. D. McCarthy, and M. Zald (eds.). Cambridge: Cambridge University Press.

Mahdavy, Hossein, 1970. "Patterns and Problems of Economic Development in Rentier States: The Case of Iran," *Studies in the Economic History of the Middle East: From the Rise of Islam to the Present Day,* M.A. Cook, (ed.). NY: Oxford University Press.

Marshall, Monty G. and Ted Robert Gurr, 2005. *Peace and Conflict 2005: A Global Survey of Armed Conflicts, Democracy, and Self-Determination Movements.* College Park, MD: Center for International Development and Conflict Management.

Marshall, Monty G. and Keith Jaggers, 2000. *Polity IV Project: Dataset Users Manual.* College Park, MD: Center for International Development and Conflict Management.

Most, B. A. and H. Starr, 1989. *Inquiry, Logic and International Politics.* Columbia, SC: University of South Carolina Press.

Political Instability Task Force, 2003. *Political Instability Task Force Phase IV Findings.* MacLean, VA: Science Applications International Corporation.

Reno, William, 1999. *Warlord Politics and African States.* Boulder: Lynne Rienner.

Rotberg, Robert I., 2003. "Failed States, Collapsed States, Weak States: Causes and Indicators," *State Failure and State Weakness in a Time of Terror,* Robert I. Rotberg (ed.). Washington, D.C., Brookings Institution Press, 1-28.

——, 2004. "The Failure and Collapse of Nation-States: Breakdown, Collapse and Repair," *When States Fail: Causes and Consequences,* Robert I. Rotberg (ed.). Princeton, NJ: Princeton University Press.

Rubin, Barnett, 2002. *The Fragmentation of Afghanistan: State Formation and Collapse in the International System 2nd edition.* New Haven: Yale University Press.

Simon, Marc V. and Harvey Starr, 1996. "Extraction, Allocation, and the Rise and Decline of States: A Simulation Analysis of Two-Level Security Management," *Journal of Conflict Resolution,* 40:2, 272-297.

——, 2000. "Two-Level Security Management and the Prospects for New Democracies," *International Studies Quarterly,* 44, 391-422

——, 2002. "Substitutability and the Internal-External Conflict Linkage: Modeling Policy Trade-Offs Between Security, Resources, and Strategy," Paper presented Annual Meeting of the American Political Science Association, Boston MA, Aug 29-Sept 1.

Snyder, Richard, 2006. "Does Lootable Wealth Breed Disorder?" *Comparative Political Studies,* 39:8, 943-968.

Steinbrunner, J. D., 1974. *The Cybernetic Theory of Decision.* Princeton, NJ: Princeton University Press.

State Failure Task Force, 2000. *State Failure Task Force Report: Phase III Findings.* McLean, VA: Science Applications International Corporation.

Thomas, George M, John W. Meyer, Francisco O. Ramirez and John Boli, 1987. *Institutional Structure: Constituting State, Society and Culture* (eds.). Beverly Hill: Sage.

Tilly, Charles, 1975. "Reflections on the History of European State-Making," *The Formation of National States in Western Europe,* C. Tilly (ed.). Princeton, NJ: Princeton University Press.

——, 1978. *From Mobilization to Revolution.* Reading, MA: Addison-Wesley.

——, 1992. *Coercion, Capital and European States.* Cambridge, MA: Blackwell.

Vandewalle, Dirk, 1997. "Islam in Algeria: Religion, Culture and Opposition in a Rentier State," *Political Islam: Revolution, Radicalism or Reform?* John Esposito (ed.). Boulder: Lynne Rienner.

Walt, S. M., 1987. *The Origins of Alliances.* Ithaca: Cornell University Press.

Weber, Max, 1949. *The Methodology of the Social Sciences.* NY: The Free Press.

World Bank, 1997. *World Development Report 1997: The State in a Changing World.* New York, NY: Oxford University Press.

Appendix 1: Formulas and Baseline Initial Values for Simulation Parameters

Symbol	*Definition*	*Initial Value*	*Formula/Rule for change*
C_i, S_i	Government i's defense *capabilities* against external threats, or domestic *strength* against internal threats	100 for major power, 50 for minor power	
SOC	Societal resource base	100 for major power, 50 for minor power	Grows by 1% after each round
$ECap_i$	State i's Extraction Capacity	1.0 for normal state, 0.5 for failing state	$C+S_{current}/C+S_{initial}$, (excludes resources extracted from ally)
c_i	Government i's capabilities compared to other state j		$c_i=1-[C_i/(C_i + C_j)]$
s_i	Government i's strength compared to rebel group k		$s_i=1-[S_i/(S_i + S_k)]$
S_k	rebel group strength	40 in major powers, 20 in minor powers; 10 for militia groups	Grows by 2% each time government extracts (1% for militias). Militias also get 10% of what the government extracts.
p_j	interaction opportunity	0.7 between major power and all others, 0.6 between minor power and all others;	Increases by $0.03(1-p_j)$ after ally extraction
w_j	Perceived *willingness* of another state to attack	0.10 (major to major) 0.06 (minor to minor) 0.04 (minor to major) 0.07 (major to minor)	Reduced by 3% after an allocation.
r_k	Perceived *willingness* of rebels to intensify opposition	0.06	Reduced by 3% after an allocation from Gov't i.
R_i,T_i	External *Risk* to government i from other state j; Internal *Threat* to government i from rebel group k		$R_i = p_j w_j c_j$, $T_i = p_k r_k s_k$
H_i	frequency of extraction vs. allocation—external threat	50% of rounds	Grows by 0.03(1-H) or falls 3% based on results of last 5 rounds
HH_i	frequency of extraction vs. allocation—internal threat	50% of rounds	Grows by 0.03(1-HH) or falls 3% based on results of last 5 rounds
G_i	legitimacy	0.00 (neutral)	After extraction, rebels grow (or shrink, if G<0) an additional G*SOC/S/T or G*SOC/C/R

Index